What Will It Be Like To Live On Mars?

- The sky will be pink.
- There will be absolutely no smoking—the air will be so combustible that even striking a match will be forbidden.
- Birthdays will come around only every two years, since the Martian year is 687 days long.
- Summers will be six months long, but the nights will be bitterly cold.
- Martian voices will fall to a deep growl, due to the extra carbon dioxide in the atmosphere. Soprano singers will be highly prized.
- Though people will live in hermetically sealed houses, eventually only a light oxygen mask will be necessary for a walk in the breathtaking Martian landscape.
- Because the air will be thin and the gravity far less forceful than on earth, human colonists could develop strong lungs and weak legs.
- Eventually Martians will become a completely different species, unable to interbreed with the Terrans (humans) who were their forebears!

James Lovelock, praised as "an extraordinary British scientist" and an "inventor of distinction," is a Fellow of the Royal Society and a former specialist with NASA. Mr. Lovelock contributed significantly to the design of miniaturized gas chromatographs, two of which flew to Mars on the Viking landers in 1976. He also invented a tiny electron-capture detector, currently in use in antiterrorist devices.

Michael Allaby, the distinguished editor of the Oxford Dictionary of Natural History, is a highly respected science writer and the author of more than twenty books.

The Greening of Mars

**Michael Allaby
and James Lovelock**

WARNER BOOKS

A Warner Communications Company

Contents

Introduction

The ground slopes gently upward to the horizon, where the tall, green plants give way to a small wood, mainly of tall horsetails. A thin wind rustles the vegetation, but is strong enough to cause only the slightest swaying. Here and there, where a rock crops out through the soil, there is a splash of bright colour where lichen have begun the process that eventually will crumble the stone on which they grow. Apart from the faint rustling, the only sound is made by the rush of a small stream, nearby but out of sight, as it tumbles and rolls the loose pebbles of its bed. A white cloud moves slowly across the pale pink sky.

A man comes into view, walking quickly as though on an important errand. He is dressed casually in light clothes appropriate for a warm summer day, but his face is partly concealed by the breathing mask he wears to supply him with oxygen from a small pack on his back. It seems I have a visitor.

As I watch from my window, idly, yet preparing to be

interrupted in my idling, I am struck again by the beauty, the serenity of Zephyria, and by the contrast between my life here and the almost frenzied activity to which the inhabitants of Earth condemn themselves.

The man has turned a little to one side. He will pass by my house on an errand that takes him to disturb someone else. I am pleased.

This record I have just completed describes the human colonization of the planet Mars and the ways in which humans have learned to live here. As its story develops my reasons for writing it are made clear. If it comes to be read on Earth, and it is not intended primarily for Terrans, perhaps it may rekindle in them a spirit of adventure they seem to me to be in danger of losing. It will remind them of that past we have shared.

Perhaps it should have been written in the troubled years of the terran twentieth century, a few years before the great adventure began. Then it could have stood as a prospectus, perhaps, a little like that which Columbus prepared to impress those who might fund his voyage to the west. Would it have been possible then, before the somewhat haphazard operations began, to assess the scientific and technical tasks that would be involved in such colonization? It should have been. All the necessary ingredients were there for those able to perceive them. It should have been obvious that the transformation of that dead planet into a place in which living organisms from Earth may thrive would be much easier than had been supposed.

Even the cost of the venture was not exorbitant. In fact it was met largely by the colonists themselves, for our colonization programme owed little to governmental aid. It was, and had to be, a private undertaking.

Knowing that, of course, would have made it no less obvious that the human settlement of Mars was also imminent. Historically, it seems reasonable to allow that when humans are equipped to do something, before very long they will do it, and not only were they equipped as long ago as the 1980s, but all the equipment could be obtained by those who needed it without too much difficulty. Even the vehicles needed to transport people and equipment could be provided, although such elaborate ships were not required at first.

A curious series of historical coincidences had led to a situation in which Mars could be colonized using the technologies and materials that were available, and with several important political benefits to the people of Earth.

In going through the notes I made, mainly on board the *Arcturus*, I had to decide how to tell the story. I could have attempted a rigorous intellectual description of the science, the technology, the politics. That would have demanded close reasoning, supported by the facts, all listed in footnotes and references. It might be possible, but it would be dull, and for that reason perhaps unsuitable for a description of the greatest of all human adventures.

I chose, therefore, to tell the story as it happened. I had little reference material beside me as I recalled what I had been taught of our history, but my notes included my thoughts at the time they were made, and had some spontaneity. So I did not disappear for long periods into the archives. Instead I merely expanded my notes, and so was able to tell the story not only of our planet, but of my voyage as well. If this makes the account anecdotal, and for this reason academically 'disreputable' it is the price I must pay.

I did devote considerable attention to an aspect of the theory behind the reformation of the planet that fascinates me. In essence, this maintains that since their first appearance on Earth, living organisms have taken over the 'management' of the planet, regulating its surface temperature and its chemical cycles for their own benefit. This being so, terran scientists realized that the transformation of Mars might be handed over to similar organisms, provided they could be given a little help at first. This help took the form of certain chemicals which, injected into the martian atmosphere, began to raise the atmospheric temperature. The carbon dioxide that formed a hoar frost in winter sublimed and remained in the gaseous phase, and some water evaporated. From this point, living organisms took over—but I must not fall into the trap of telling my story twice!

Where it is necessary to understand a scientific concept in order to understand the story, the concept is explained, I hope, clearly and without the use of jargon.

There is one convention I do not explain. Indeed, although I can imagine why the words it uses were chosen, I have no idea when it began. As soon as people began to leave Earth, it became necessary to distinguish between the things of Earth and the things of Mars, and adjectives were required. 'Martian' presented no difficulty, written with a capital letter if it referred to people, as a noun, and with a small letter if it did not, as an adjective, but Earth did. 'Terrestrial' would have been misleading, since it is used now, as it was centuries ago, to distinguish those things associated with land from those associated with water. 'Earthling' might have described a human, but only a human, and it makes that human sound retarded. So the word 'terran' was coined, and today we use it in the same way as we use 'martian'.

Read on, and as you read try to imagine you are a Terran of the twentieth century. Try looking upon this real history as though it had been written before it took place. Perhaps you will understand then just how extraordinary talk of migration from Earth must have seemed to ordinary people. Then go on to remember that in the middle of the twentieth century some solemn, educated people were saying that humans could never travel in space. Not long before that it was said that vehicles could not fly that were heavier than air. Not long before that there were those who believed that humans would die if they attempted to travel faster than thirty miles an hour.

I mention this partly to emphasize the dimensions of our achievement, and partly to warn you to give due consideration to plans that may be made before much longer to explore more deeply in space. Do not dismiss as impossible attainments that may be closer than you imagine. Uninhabitable Mars was induced to support terran life. Today humans walk across martian landscapes without wearing spacesuits. It happened, and it may have been only a beginning.

Launch

The woman in front of me was very nervous. Waiting in the departure lounge, while others stood or sat around in small groups talking, or reading magazines, she paced. She would walk up to a travel poster, stare at it with exaggerated absorption, then move away after a moment, probably without having seen the poster at all. She could not rest, and she was alone.

When our launch was called, and we walked into the open air, she strode as though impatient, then stopped as she, as we all, saw for the first time, in the distance, the vehicle that would carry us into orbit; brilliantly white in the sunshine, upright, a temporary visitor to Earth, in a hurry to be gone.

I sympathized with her, yet felt superior. Few people make this journey twice, and this was my second voyage. I was seasoned, a veteran. My first journey had ended here, at this spaceport, two months ago. It was hardly time enough to grow acclimatized. The land was beautiful, the culture exciting, all much as I had expected it to be, but the visit had left me tired. While she walked a little stiffly, I laboured, heavily.

The trouble was that although every effort was made to prepare people for the journey, they arrived unprepared at their point of departure. You cannot prepare people for an experience that is unique. It is like trying to prepare people for death, or for their first love affair. You would watch the simulations they sent with the tickets, but no matter how often you watched them, with no matter what intensity of concentration, you saw but pictures, heard only sounds, objective where experience is subjective.

In their interviews with the selectors all these people, including the nervous woman, would have recounted intimate details of their life histories, of their beliefs, hopes and fears, and have been told all the selector could tell them in the time available, and still they were unprepared. It is all abstraction, all theory. The simulator can present sights and sounds, accurate enough in themselves, but we perceive the world with more than eyes and ears. It does not, cannot, convey the faintly perfumed smell of the air conditioning in the departure lounge, the smell of the bus, or of the launch vehicle itself. It tells nothing of the suddenly cool air, the brightness of the light, as you step from the lounge. Even its sounds are incomplete. Where, on either of the two sound tracks on the disc, will you hear the slight rattle caused by a minor mechanical fault in the engine of the bus, or the sudden cry through a moment of stillness of the small child whose cuddly toy falls to the ground? These are details, wholly irrelevant, but it is from small, irrelevant details that we know the world.

Perhaps the preparations miss the point. Perhaps they would arm us better were they to deal only with irrelevancies. After all, we may trust the crew to operate the equipment, to take care of us. For us, perhaps it is the wealth of technical detail with which the simulations are packed that is irrelevant. We need only the reassurance that comes with familiarity.

Most of the travellers, even today, are migrants. They are latecomers, of course. It is their ancestors who were the true spiritual descendants of the waves of ambitious, romantic, impoverished, oppressed, frustrated voyagers who crossed the oceans of Earth centuries ago to build new lives in new places. They even called the Americas the 'New World'. What would they have thought of a journey through space, to a world that was hardly new, but that was most certainly a world, complete unto itself? I dare say the mode of transport would have startled them, but the travellers would not. They would have known them, would have identified with them, would have found much to discuss concerning dreams and nightmares, new lives and old.

In the early days attempts were made to restrict immigration. Facilities had to be provided by the colonists themselves, after they had arrived, and it was believed to be important, perhaps it really was important, that those who came to Mars should have special, needed skills. We needed people who could be construction engineers and farmers, teachers and doctors, biologists, geologists, and people with genius in their hands who could make familiar, everyday things without which life is impossible, by improvised techniques from such raw materials as the long-dead planet might supply. Most especially we needed people who were adaptable enough to turn their hands to anything. We needed people with special skills, but we did not want narrow specialists jealous of their professional boundaries. Our museums fill quickly, for in the 'stone-ceramic-and-metal-age' everything was made from those materials, and such things have lasted. Newcomers are surprised to see tables and chairs made from ceramics, but why not? We needed cyberneticists, too, the students and manipulators of systems. It was they who found ways to make the planet habitable. Eventually the restrictions were relaxed. By then we needed people, to increase our numbers.

That is not to say that the party now walking in a long line from the bus to the elevator lounge had not been selected at all. Each of them had passed a medical examination. Our Council insisted on that, for we had no wish to import diseases or genetic imperfections that were not present among us already. I had begun to question the wisdom of that selection. The report I carried, sealed under a security lock in the shoulder bag that was integrated with my coat so it could not be removed, described the outcome of too rigid a separation of peoples under novel circumstances. If its conclusions were true, as many of the finest scientists on both planets believed them to be, relations between the peoples of the two worlds might be more difficult in future.

Not that genetic selection was entirely to blame. We could have remedied imperfections, after all, as we had remedied them among ourselves and as we continued to remedy them in those immigrants who slipped through the net despite our precautions. It was costly, but not too difficult. Our own preferences, the images of beauty we had developed over the generations among ourselves, had contributed at least as much to the dilemma in which we now found ourselves.

In any case, no matter what we might decide on Mars, some form of medical selection would continue because the space-lines demanded it. They required the bodies they carried to be fit enough to survive the journey. Essentially, this meant that only passengers with sound heart and circulation could be carried, for no matter how sophisticated the technology might become there was no way to reduce, far less to eliminate, the physical stress caused by a rapid acceleration from the surface of Earth to orbit, and the acceleration had to be rapid to conserve fuel. I imagine we would accept human corpses on Mars. On our planet they are rather valuable for the chemical compounds from which they are made. It is the spacelines who object. They believe it is bad for crew morale and even worse for business to sell tickets to living customers and to deliver dead ones. You can see their point.

We did introduce elementary psychological checks on prospective immigrants. You can compare our immigrants to those who left Europe for the Americas and Australasia. The journey and its implications are similar, but only partly so. America and Australasia were strange places in the early days of their colonization, but they did possess soil, water, plants, and animals, resources with which the immigrants could work, and environments into which the introduction of their more familiar crop plants and livestock was fairly easy. What is more, the return journey was possible, at least theoretically, for those who could not settle to their new lives. Contact with 'home' was not broken entirely. Conditions on Mars were utterly different, at least to start with, from those on Earth. Homesickness could and did develop rapidly into severe mental illness, and the return journey to Earth was a great deal more arduous and, more important, far more expensive than a sea voyage half way round Earth.

Even then, as I am sure you know, America and Australia had their social problems during the pioneer days. One of the first Martian Councils debated whether we wanted, or could afford, a 'wild west' on the planet. We might not be able to avoid having one, but it seemed foolish to volunteer for it. So those who would migrate were interviewed before they were accepted by Terrans working under the supervision of our own people, the selectors, who had direct experience of the problems the new arrival would encounter, and who were presumed to know the needs of the planet. Those checks continue, but today they are little more than a formality. Life is easier, and in fact a 'wild west' never did develop.

You go from the bus to the elevator lounge, a carpeted but unfurnished room that is just large enough to contain all the passengers, together with the stewards who accompany them. From there, ten at a time, you enter the elevator that carries you up to the entry port. Everyone knows about the elevator. What you may not know is that it travels very fast indeed. It must, because the passengers must be embarked as quickly as possible but in small batches, and there is more to embarkation than just climbing aboard and finding a seat. At the medical examination, each passenger sits in a device that makes a mould of the back of the body, from head to heels. This mould is used to make a seat lining contoured precisely to the body, which has to be fitted to the seat in the vehicle. It means that each passenger must be shown to a particular seat, then fitted into it and strapped down, and hand luggage must be stowed—except for my shoulder bag which remained with me by diplomatic privilege. Passengers can be embarked only in small batches to avoid crowding and confusion in the cabin, and while it is all going on the vehicle sits waiting, fuelled and devouring money by the second. I suppose that in the general anxiety about wasted time someone decided the elevator should not be permitted to cause delays. So it feels like an appropriate prelude to the launch itself as it hurls you upward. They say nothing about that in the video. My nervous fellow traveller was thrown off balance, clutched frantically at my arm to save herself, became embarrassed at such close physical contact and blushed, then stared at me in silent, wide-eyed terror before lurching away from me as the elevator stopped abruptly and she was thrown again.

We found ourselves seated side by side, but we were among the last elevator batch to board and there was time only to introduce ourselves before the launch itself interrupted our conversation. Her name was Towers, she told me, and she was being sent to Mars by the corporation for which she worked. I suspected it was a promotion-and-posting of the kind many organizations use to rid themselves of personnel who have offended or caused embarrassment. The practice is hallowed by terran tradition and it would take more than our martian selectors to put an end to it, even had we wanted to do so. In fact I doubt that we would. The offenders, the gauche, often turn out to be the people of imagination in an ossified institution, and we are the beneficiaries. So Towers had passed the immigration tests, but her journey was not entirely voluntary. It was sad for her, but not too serious, for we take trouble to make such people feel wanted. At a more personal level I looked forward to improving our acquaintance, for beneath the nervousness I thought I detected a lively, witty personality.

What is more worrying is that people like Towers, who have not planned and dreamed for years of starting life anew, know very little about our planet and our history. The information is available. You can find it if you wish, and what I may call the 'deliberate' immigrants will have found it. No doubt many of those sitting in the vehicle knew more about Mars than I do, who was born here. But Towers set out knowing almost nothing. Only the barest details are taught in terran schools, absorbed as they are in the long, intricate history of their own world. Terrans who leave, like Europeans who left for the colonies all those centuries ago, are largely forgotten by those they leave behind them. They cease to be part of the continuing history of their original country or planet. They disappear from the picture.

Even what is taught is distorted, or so it seems to us. Individuals who lived on Earth and whom we regard as heroes, founding fathers if you like, whose contributions to the beginning of our history are quite fundamental, are largely ignored. Claims to a place in terran history must base themselves on contributions made on Earth and to Earth. In any martian school their story is told differently.

There was no time to talk as we were ejected into orbit, and the brief time we spent above Earth before docking with the spaceship was too precious to permit chatter. That is the time for passengers to gaze back upon Earth, which is now above their heads because of the attitude of the vehicle, and visible through windows. They gazed in silence, and so did I, at the most beautiful sight most of us will ever see. Then the vision disappeared as we entered the ship.

On the ship, though, as we sailed through the solar system, there was time to make notes, to give Towers some insight into the world that was to be her home. At first, I imagined I might present them to her as we landed, a bundle of papers for her to read at her leisure. I felt I owed it to that startled woman who looked as though she had lost control of her life. Perhaps I owed it to myself. In the event, it was not to be. Conversations I had with her attracted others, and before long they developed into a series of informal discussions which I led from notes I prepared in advance. Perhaps, then, those notes, with a minimum of editing and including my own reflections as I wrote them, may serve others. Who knows but that someone, at some time, may see fit to publish them as a brief history of our planet.

Towers is important to my story, for in a sense it exists because of her, but she is not a 'character' in it, and she will not appear in it again.

Only where should I begin? I had been thinking about people, so perhaps my story should start with Sir Travers Foxe. And yet . . .

We honour him, as we honour the great and good men and women in our past, and most of us would accept that without him our history would not have begun. Yet it cannot be so, not entirely. Far be it from me to minimize the contribution he made, but had it not been him, might it not have been someone else? If history had not unfolded as it did, might it not have unfolded in a rather similar fashion nevertheless, so that today we would remember different names, slightly different dates perhaps, but events that were not very different?

Events, and the ideas that give rise to them, have their time. They emerge, almost unbidden, from the circumstances of their age, and of the ages that went before, with a sort of inevitability. They could not happen at another time (the internal combustion engine could not have been invented in the fourteenth century, for example, nor in the twenty-third). The people, the individuals who 'make' history are but details, instruments to be used as events take their course. Many people have suspected as much. It is this thought that gave rise to so many determinist theories of history, and to the spurious 'art' of fortune-telling. Today we know there is nothing magical about historical processes. We, and the ways in which we interact with one another and with the world around us, comprise a dynamic system, and the system can be modelled. Soothsaying can be reduced to computer forecasting, and these days the technique is reasonably accurate. We can foretell the future, sometimes for as much as a week ahead.

But I digress. The point I was trying to make is that before we consider the role of individuals in history we must examine the times in which they lived. 'Great men' do not 'make' history. History produces circumstances in which they, and their actions, are bound to occur. Their time is right. We must return, then, in our imaginations, to the final quarter of the twentieth century.

Political power, you may recall, resided in the governments of nation-states, of which there were many. Among them, two were pre-eminent, the United States of America (USA) and the Union of Soviet Socialist Republics (USSR). These two had emerged from the Second World War (1939-45) occupying or otherwise dominating a large part of the inhabited land surface of Earth, so constraining many smaller national governments. Having been victorious in that war, their military leaders, and the industries manufacturing military equipment, were numerous, strong, and healthy. The two were rivals and expressed their rivalry mainly by the accumulation of armaments.

The 'arms race' consisted in the acquisition of weapons no one seriously intended to use. It operated psychologically by exploiting ancient fears. The USSR had been invaded many times in its history, most recently in the 1940s, and its people had a deep-seated dread of war fought on their soil. Traditionally, their response to this dread had been to drive out the invader, establish a new frontier, and settle the land behind that frontier. It is how their territories were expanded from Europe to the Pacific. Europeans were encouraged to see this as simple expansionism, and it preyed on their fear of war, but also of invasion from the east. Russians considered themselves Europeans, but western Europeans had long memories of the Mongol invasions, from central Asia, centuries earlier, and they did not distinguish clearly between Russians and central Asians. Each sought to protect itself from the other, and interpreted such essentially defensive moves as threat.

The world was divided, then, but not without protest, for it was an age of popular protest and the protests often succeeded. In the 1950s anti-nuclear campaigners demonstrated in favour of a ban on the atmospheric testing of nuclear weapons. Such a ban was signed, although a few non-signatory nations ignored it. The protest then adopted civil rights as its cause in the USA, achieving substantial reforms, after which it turned to opposition to the war then being fought in Vietnam. The war was terminated by the withdrawal of US troops. Partly overlapping with this movement there grew up another, in most of the western countries, urging immediate action for the protection of the natural environment. This was even more successful than its predecessors. The United Nations formed an environmental agency, almost all national governments appointed ministers of the environment, environmental problems were scrutinized at a high scientific and political level, and the condition of the natural environment began to improve.

It was at this stage, in the early 1980s, that protest returned to the arms race, but now the protest movement had gained experience, was flushed with earlier successes, and it became very influential very rapidly. It demanded principally an end to the stockpiling of nuclear weapons, and with the support of many church leaders to supply moral force it compelled governments to listen to it.

They listened, but they could afford to ignore many of its arguments, because, like many of the more extreme environmental arguments before them, they were flawed by errors of fact. The protesters maintained that the use of weapons of indiscriminate mass destruction was immoral, and that therefore the holding of such weapons and the threat to use them, which was the basis of the deterrence policy of the time, was similarly immoral. It assumed, indeed stated from time to time, that nuclear weapons, uniquely among weapons, were weapons of mass destruction. In the 1950s and 1960s that had been true. The weapons were meant to produce large explosions and to do so at great distances, carried by rockets or aircraft, and rockets in particular could not be guided to their targets accurately. By the 1980s, however, guidance systems had advanced and the size of the weapons had diminished in consequence. Nuclear weapons were no longer necessarily weapons of mass destruction and the time was fast approaching when they could be directed over thousands of kilometres to individual buildings or installations. They could be confined to military targets, civilian populations might be spared, and the explosive devices themselves were 'cleaner', in that they produced much less radiation than had those from which they were developed.

There were, then, three forces at work that were crucial to the events which followed. Led by the USA and the USSR, the nations of Europe were divided by fear, and both major powers were engaged in an arms race that was accelerating. The arms race was opposed vehemently by a popular movement drawing support from many of the most respected figures in society. The environmental movement had achieved its objectives so effectively that by the 1980s the environmental consequences of any proposed development had to be taken into account before that development could be permitted to proceed. 'Environmentalism' was fashionable, accorded a high priority in the taking of decisions, and projects promising to enhance the quality of

the environment, however that might be defined, were
likely to win support from governments.

When things began to happen they happened in rapid
succession. By the end of 1982 the western nations were
planning to deploy new weapons systems and to replace
certain older ones with newer versions. This provoked
strong opposition from those who began to demand the
general disarmament that before they had only requested.
Their demands went largely unheard, because their arguments could be ignored.

I mentioned earlier the confusion over weapons of mass
destruction. This weakness in the argument was made
worse by extreme statements made by some of the would-be
disarmers. They claimed that a full-scale war, using all the
weapons then in existence, would bring mankind to extinction and might well render the entire planet uninhabitable
for life of any kind. There were many uncertainties in the
political and military situation of those times, but this was
one issue for which facts could be supplied with some
confidence. Such extreme claims were preposterous, and
politicians knew it.

Two atomic bombs were used at the end of the Second
World War and the effects had been studied in great detail.
The sites in the Pacific used for atmospheric weapon
testing before such tests were abandoned were also studied, and the rate of their natural recolonization by plants
and animals was monitored. From such studies as these,
backed by more theoretical assessments, it had been calculated that a full scale war would cause appalling devasta-
tion and loss of life in the combatant countries, but that
even in those countries there would be survivors, probably
amounting to about half the pre-war population. In non-
combatant countries effects would be minimal. Natural
recolonization would ensure that after about thirty years it

would be difficult to find biological evidence that the war had taken place at all—though ruined buildings would remain.

Further support for this view was supplied as evidence accumulated to explain the mass extinction of species that had occurred about sixty-five million years earlier. This, it transpired, was due to the impact on Earth of a planetismal, releasing energy many times greater than the entire arsenal of weapons available in the early 1980s, yet the planet had not been incinerated. Life had survived, as it had survived many other such impacts in its past, all of them before the earliest recorded human history. Clearly, survival was not just possible, it was certain.

It was in the mid 1980s that the leaders of the peace movement changed their position. They ceased to claim that nuclear weapons were uniquely weapons of mass destruction, and they dissociated themselves from the idea that mankind would not survive a nuclear war. Instead, they returned to the morality of the use, and threat of use, of weapons of mass destruction, but this time concentrating not on the weapons but on human behaviour. The possession of a physical artefact is morally neutral, after all. It is the use made of it that can be weighed morally. They demanded not that governments dispose of their weapons, but that they renounce the indiscriminate use of those weapons against civilian populations. If accepted, this would require the removal of all cities, conurbations and other population centers from lists of targets. Since most of the older, larger weapons were directed against such targets the new demands were rather similar to the old in their implications, but they were on much firmer moral ground.

The demands could be ignored, but if governments replied to them they were compelled to declare whether or not they would sanction attacks against unarmed civilians, and whether they would regard such attacks as an appropriate response to a military threat. They were placed in a difficult dilemma, and the influence of the peace movement grew. Long before 1990 no political party in any western country could expect to be elected unless it made firm promises to disarm, and no government could expect to be re-elected unless during its term of office it had made real progress toward disarmament.

Armament continued for a time, and much of the promised replacement of weapons took place during the latter part of 1983 and 1984. It was only later that general disarmament began in earnest, and stocks were reduced.

Everyone took an interest in disarmament, and almost everyone concentrated that interest on the warheads—the weapons themselves. There were few people who bothered about what was happening to the obsolete delivery systems—the rockets.

It is time to return to Sir Travers Foxe, for he was one of those few who speculated about the fate of the abandoned rockets.

He was born in Dundee in 1923, as plain Graeme Foxe. He added his mother's family name 'Travers' later, when he became better known. He trained as a civil engineer, developed an interest in town planning, and at the age of thirty, feeling his apprenticeship was complete, he joined with a few colleagues to form a partnership that was responsible for a number of urban developments that were much admired. He always sought to improve the man-made and natural environments in the areas he redesigned and in time he came to be respected widely.

As his firm won more overseas contracts and he began to travel the world, so he grew friendly with many leaders of the countries in which he worked, and he exploited those friendships in the promotion of harmony. He shared the popular concern about the imminence of war and it was said that more than once he arranged private meetings at which disputes were resolved that might otherwise have led to conflict. He came to be known as a peacemaker as well as a sensitive planner.

He was invited by the United Nations Environment Programme to advise on urban developments in several African and Asian countries, and for some years he found himself based partly in Nairobi, at UNEP headquarters, and partly at the UN Building in New York.

Foxe was also concerned, privately, about Mars. The first landings on the planet were made by the Viking I and II vehicles in July and September, 1976. They excited the media briefly, but interest soon waned and as western countries entered the recession of the 1980s, thoughts of further excursions to Mars receded. Yet to Foxe those first landings uttered a warning that seemed to grow stronger as the years passed, for the thoughts that came to obsess him must have occurred to others. The planet had become accessible and sooner or later it would be exploited. It needed official protection, and what better time could there be to protect it than one in which no one planned to visit it again and therefore no one was likely to object?

He began to sound out his UN colleagues, some of whom were high officials, and together they devised a scheme that might work. In order to work it must cost nothing, at least to the UN itself, for UNEP was suffering already from the failure of certain member nations to pay their agreed subscriptions and any suggestion that the UN grow still larger, and spend still more money, would meet strenuous opposition. The General Assembly passed without demur a resolution that entrusted UNEP with responsibility for safeguarding Mars, on the clear understanding that the cost of carrying out that responsibility would not entail any increase in the UNEP budget. It was little more than a formality, for Mars was remote and the problems of Earth were immediate. Quietly, without fuss, Foxe and his UN colleagues recruited a small group of retired politicians and industrialists, who formed a consortium, and UNEP transferred the administration of Mars to them. The duties were not onerous. Activities on the planet were to be prohibited if they threatened injury to the environment there, no military installations were to be permitted, and if the resources of the planet should be exploited, they had to be exploited for the benefit of mankind as a whole. The consortium had the task of examining all applications to visit Mars, and of issuing licences to do so. It could charge for the licences it issued, and could award franchises for permitted activities. The proceeds of these charges were to be passed to UNEP, after the handling costs had been deducted by the consortium. However, the consortium was also empowered to make its own investment in projects of which it approved. The capital for such investment could be taken from the UNEP share of the income from the sale of licences and franchises.

Once the resolution had been passed, the necessary papers composed and signed, and the consortium was formed and working, everyone accepted that the matter had been resolved most satisfactorily. No one noticed that the wording of the clause on protection of the martian environment was open to two interpretations. It could be held that the environment must not be disturbed, that it must remain in its pristine, arid state for all time. It could also be held that its enhancement as an abode for humans and other species from Earth was not inconsistent with its protection. Nor did anyone point out that the agreement

made Sir Travers Foxe to all intents and purposes the
owner of the planet.

And that is why we regard Foxe as our founder.

Nothing happened for a few years, and then Foxe's concern for the environment manifested itself again, this time
back on Earth.

As I mentioned, he was worried about the rockets that
would be abandoned as the weapons systems were mod-
ernized in the mid-1980s. The rockets were of two types,
liquid-fuelled and solid-fuelled, and it was the solid-fuelled
ones that concerned him, for they were inherently dangerous. The fuel itself consisted of a combustible material
formed into a plastic, rubbery consistency. Molten, it was
poured into a rocket casing, where it set. The inside of the
casing, and so the fuel, had a complex shape designed to
ensure a particular, desired rate of combustion, but it
meant that once it had solidified within the nooks and
crannies of its housing, the fuel could not be removed.

There were several thousand such rockets, with resound-
ing names such as Poseidon, Polaris, and Trident. The
USA alone had about four thousand of them, and Britain
and some other countries had a few more. Foxe imagined
them lying in some store, in the countries that owned
them, but most of them in the USA, behind doors locked
with locks that would corrode and break, protected by
guards who would grow lax and eventually be withdrawn.
So far as anyone knew, the fuel was stable and would
remain so, but how much did anyone know? The situation
was novel, but were there to be an accident it could be a
serious one.

In those days it was easy enough to generate alarm when
the cause of that alarm was a technological device. Foxe
mentioned his fears to government ministers. There was,

after all, the possibility of a press exposure that would provide the peace and environmental movements with more sticks with which to beat the backs of politicians, quite apart from the genuine risk itself. He found ministers sympathetic and, when he offered to dispose of the rockets for them, grateful.

He proposed to arrange for the collection of the unwanted rockets. He would have them removed to his own site, or at least to the site he rented cheaply in the southern part of Tamil Nadu, shipping them to Madras, then transporting them by rail to Tiruppur and by road from there to the site itself, from which they would be launched into deep space, and so lost from Earth for ever. Everyone appreciated that the scheme brought several incidental benefits. A fairly substantial amount of work would need to be done at the site. This would provide employment for several hundred people, who would be accommodated there along with their families. The work would be designed to employ local labour, which would be helpful economically, and when it finished, if it did finish, the installations could be adapted to another use. The scheme was attractive, and funds were made available for the Indian development. A contract was prepared and Foxe was made responsible for solid-fuelled rocket disposal, being paid an agreed sum for each rocket removed.

Foxe now owned Mars and several thousand rockets, but the coincidence went unremarked. Who would want to go to Mars?

As work progressed in Tamil Nadu, Foxe began to wonder whether all these rockets might not be put to some use. They were designed to carry a payload and without a weight on the nose they would be unstable in flight. So they would have to be ballasted before launching, and since they had to be ballasted, was there not some noxious material of which he could rid the world?

Of course, there were many, but most were unsuitable. Nuclear wastes, for example, were being recycled, and the final wastes from the nuclear cycle were being disposed of perfectly well, but after a bitter environmentalist campaign. It would be better not to revive that old quarrel. Others were unstable, so that after a period of storage they presented no hazard to living organisms. He needed a substance that was considered to be highly noxious, for which there was no conceivable use, and which was so stable that disposal of it was almost impossible. Before long he found his waste.

In the late 1970s the fashionable environmentalist issues tended to concern the ozone layer, a region of the upper atmosphere in which the energy of ultraviolet solar radiation causes the formation of ozone from oxygen, and where much of that ultraviolet radiation is absorbed. A series of mechanisms was proposed by which the ozone layer might be depleted leading to increased exposure of the planetary surface to the radiation, which was harmful to living cells. One by one these mechanisms were found to have little effect on the ozone layer—some even thickened it—but the most popular of them concerned a range of chemical compounds used in plastic foams, refrigeration and air conditioning systems, and as propellants in aerosol cans. The substances were known popularly by their trade name, 'freons', or more generally as chloro-fluoro-carbon (CFC) compounds. They were very stable indeed, which is why they were welcomed when first they were introduced. Being stable, they did not burn and they were not in the least poisonous. They did not harm the ozone layer, either, but by the time that had been discovered their use in most kinds of aerosol can had been banned, and they were being withdrawn from other uses as well.

The freons that had been made but not used lay in warehouses owned by the manufacturers where they occupied valuable space. They could not be moved, for disposal was not simple. They could not be dumped in any way that could permit them to enter the atmosphere. Sealed in containers they could be dumped at sea, but regulations governing the dumping of wastes at sea were being tightened, mainly at the insistence of the environmental groups who had found a new campaign, and the freon-makers had suffered enough already. They could be incinerated and so destroyed chemically, but at high temperatures and consequently high cost. So they sat there, waiting for someone to think of something.

Sir Travers Foxe entered the offices of the chemical corporation most intimately concerned with freons like a knight in shining armour. His offer, at the most modest of fees, to relieve them of an intractable problem was received with joy. The contract was signed two days later, two days being the time it took for the typing.

Foxe now owned Mars, a lot of rockets, and several warehouses full of cans of CFCs.

A conference of atmospheric scientists, meeting in Oslo in the spring of 1985, was told of the solution that had been found for the disposal of the surplus CFCs. The conference was well attended, by delegates provided with ample expense accounts, and it proceeded at a leisurely pace, to the benefit of the local hoteliers and restaurateurs. Whole days were set aside in the programme for private discussions, and tours to local places of interest. It was a most pleasant, relaxed occasion.

It was on one of those tours, on board the bus in fact,

that two chemists had a conversation that was overheard by a physicist. The physicist knew that Foxe, the new owner of the CFC stocks, also owned rockets, not to mention Mars. The chemists knew that CFCs were capable of modifying climates, provided they could be injected into the atmosphere in quantities sufficient to initiate a warming. The three of them resolved, there and then, to call upon Foxe and to propose their own scheme to him.

Red
and Dead

Until the happy conjunction of Foxe, his planetary rights, his rockets, and his chemicals brought my ancestors to the planet, Mars had been devoid of life throughout the whole of its history. It was a dead planet, a cold, dry desert beneath a pink sky. From time to time a thin wind would lift great clouds of dust and sand, producing a storm that covered a large area and lasted sometimes for weeks. Now and then a cloud would appear, white in the sky, just like a cloud on Earth, but it would pass, disperse, and it brought no rain.

People had always hoped there would be life on Mars. Indeed, there had been times when they went a long way towards convincing themselves not just that the planet was inhabited, but that it was the home of intelligent beings. In the nineteenth century an Italian astronomer, Giovanni Virginio Schiaparelli, saw what he called 'canali', long, straight lines, on the martian surface. The Italian word 'canali' means simply 'channels'; it does not imply they were made by intelligent beings. But in English the word was translated as 'canals', and canals are made by men. The 'canali' were artefacts indeed, distortions produced in the telescopes used by those who saw them—and not everyone saw them at all—and they were aided powerfully by wishful thinking. A popular picture emerged of a planet that was arid, but induced to support life by means of elaborate systems of irrigation, similar to those used in arid regions of Earth, but on a much larger scale. The impression was reinforced by seasonal changes in the appearance of the planet. Large areas would grow darker, then paler again, and it was suggested that the changes were caused by the growth of plants.

Once begun, the fantasies elaborated themselves, as fantasies will. Mars was an old planet, it was said, older than Earth, and dying. Because it was older, its life forms had been evolving for much longer than had those on Earth, and its civilizations had advanced far beyond any Earth had produced. The intelligent Martians tended to view Earth with an envious, unfriendly eye. They were, after all, inherently bellicose. Mars is red in colour when seen through a telescope, or even through powerful binoculars, and red is the colour of war, of blood, of battle standards and military uniforms. The planet was named after the Roman god of war. The idea of warlike Martians, poised to invade Earth, inspired much excellent fiction. H.G. Wells's *War of the Worlds* was the story that endured for longest. It described Mars as an ancient, dying planet, and its inhabitants evolved into a kind of helpless decadence, served by machines without which they could not survive.

The discovery, confirmed finally by the photographs

returned from the 1976 Viking landers, that Mars is red really because its soil particles are coated with iron oxides, but that the planet is devoid of life, and of any evidence of life in the past, far less of intelligent life, came as a disappointment to many people.

The popular ideas of Mars were based on scientific theories that had been abandoned long before the first landings. Better telescopes had shown there were no canals, nor anything that might reasonably be mistaken for a canal. The first close approaches by space probes returned pictures of what looked very like dry river beds, wadis, but these were far from the legendary straight lines.

Nor is Mars any older than Earth. That idea grew from an earlier theory about the formation of the solar system that was still being taught in the 1940s to eleven-year-old children conditioned to accept the statements of school teachers as indisputable fact.

According to the theory, a large object, the size of a star, passed close to our Sun while the Sun was forming and before it had fully condensed from a cloud of hot gas. The gravitational attraction of the passing star drew out a long filament of material from the Sun and as the filament cooled so it broke into globules. The globules were hot, and the rate at which they cooled depended partly on their size and partly on their distance from the Sun. The Mars globule was smaller than the Earth globule, and farther from the Sun, and so it cooled faster, and therefore was capable of supporting life sooner.

Today there is no serious dispute about the formation of the solar system. It was formed as a whole, by the condensation of a cloud of gas, dust, and galactic garbage, and at the start it was cold. As the entire mass began to rotate about its centre, tiny fragments at the edges separated to become the planets, which grew by aggregation, by collecting and holding fragments of matter that fell on to them. Everything happened together, and each part of the system is the same age as the rest. It was only after it had formed that the heating began. Pressures inside the Sun caused the start of its heating, and the decay of radioactive elements beneath the surface of some of the planets, including Earth and Mars, caused them to heat from below. There never was a hot phase during the formation of the system.

Humans are eternal optimists. Mars did not support life, and neither did Venus. That had been the other main focus of attention, and its legends included descriptions of a planet covered entirely by water, or with vast, steaming swamps supporting extraordinary monsters. Closer inspection, the penetration of its atmosphere, and finally some rather fuzzy pictures transmitted from the surface by a Soviet lander in the short time available to it before the intense heat destroyed it, revealed a world even more barren and much less hospitable than Mars. If neither planet supported life, then life must be sought elsewhere, in the galaxy beyond the confines of the solar system.

Is there life elsewhere? We do not know today, any more than our forefathers did, but even in the closing years of the twentieth century the awful suspicion was beginning to dawn that humans may be quite alone, the only intelligent beings in the galaxy, perhaps in the universe. In a sense the speculation is fruitless. There is no way to prove that other intelligent beings do not exist, but nor is there reason to suppose that the search for evidence is likely to succeed. Mathematically, the odds against making contact with another civilization are so great that, having calculated them, other civilizations may have decided not to waste their time with such projects. They may be silent.

We have moved a small distance from Earth and have established ourselves. We cannot help thinking about the colonization of other, more remote worlds. We are expansionist by nature. After all, humans have existed at all for only a million years or so. It seems inevitable that our descendants will continue the expansion, and it may proceed rapidly. It has been said that a mere thirty million years may be sufficient for humans to occupy all the habitable places in the galaxy, and our experience in Mars allows us some flexibility in our definition of the word 'habitable'. If there is life elsewhere, even if it is silent, humans may encounter it.

The debate about the existence or non-existence of 'extra-terrestrial intelligence' was summarized very elegantly back in 1983, in an article by Professor James Trefil published in the January-February issue of *The Sciences*, the journal of the New York Academy of Sciences. Trefil attacked the 'assumption of mediocrity' that lay behind the supposition that life must exist on other planets.

In the middle ages, Europeans had believed that Earth was the centre of the universe, and that human beings were by far the most important beings on Earth, fundamentally different from all other forms of life. The first shock came with the Copernican discovery that models of the solar system can be constructed in which the Sun, not Earth, lies at the centre, and that such models can account for the apparent movement of the Sun and planets as seen from Earth. Acceptance of this model relegated Earth to the status of one of nine planets orbiting a star. Later it was discovered that the Sun is a star similar to countless others. Then it was found that the solar system forms part of a galaxy, and that it lies a long way from the galactic centre, so that not even the Sun is at the centre, even of its own galaxy.

Darwin gave expression to the suspicion that was growing throughout the late eighteenth and early nineteenth centuries that humans have much more in common with other animals than had been supposed. In the twentieth century biochemists and geneticists confirmed that view emphatically. Not only do humans have much in common with other animals, to some of which they are related very closely indeed, but all life on Earth is based on similar chemistry and all species are related.

Scientists from other disciplines contributed to the work of astronomers and biologists until eventually Earth and humans had been demoted from their ancient eminence. Even the changed view of the origin of the solar system helped. An encounter between two stars close enough to produce the 'filament' would be a very rare event. The condensation of a cloud, according to known laws, must be a commonplace event.

Thus mankind emerged as but one species among millions, occupying a very ordinary kind of planet, orbiting a very ordinary kind of star, in a not very distinguished place in a very ordinary galaxy. There was nothing special about Earth, and consequently the conditions found there must be duplicated many times among the billions of stars in the galaxy, and among the billions of galaxies.

If intelligent life exists on Earth, the laws of probability dictate that it must also exist in many places elsewhere. This view was questioned, so it is said, by the physicist Enrico Fermi, who asked, 'Where is everybody?' If the galaxy teems with intelligent life, and if, as some people have suggested, it will take only thirty million years for humans to colonize all habitable sites in the galaxy, then someone should have visited Earth. Indeed, the solar system should have been colonized by at least one, and probably more than one, race of intelligent aliens. If we

can go there, then they can come here, and since the solar
system is only about five billion years old and some other
stars are much older, then there should be civilizations
with a long lead over us. They should have been here long
ago.

Attempts were made to reply to this criticism, but they
were not very convincing. Some suggested that advanced
civilizations tend to destroy themselves, so that cultures
never advance beyond a certain state and are never capable
of expanding outside their own star systems. Others fell
back on the 'silent aliens' argument, suggesting that other
intelligent beings prefer to keep themselves to themselves.
The difficulty with all such arguments is that they require
the same kind of behaviour from every single civilization.
It would take only one to break the pattern and their
presence would be revealed. What is more, the only
intelligent beings of which we have any knowledge at
all—ourselves—do not behave in this way. We leave our
home planet the very moment we are capable of doing so.

There was one source of evidence most people overlooked.
It seems very likely that any intelligent beings who wished
to study Earth would establish a base for themselves on the
Moon. It is an obvious thing to do, for many reasons.
They could unpack their equipment, spread themselves
over as large an area as they wished, cease to pay constant
attention to the condition of the orbit of their craft, forget
any fears they might have of premature discovery by
Terrans who might prove unfriendly, and enjoy an excellent view. If aliens had visited, then they might have left
relics of their visit on the Moon. The Moon provides
surface conditions that preserve artefacts and impressions
for millions of years. The footprints made by the first
terran astronauts, and the flag they raised, will still be
there long into the future, unless a meteorite happens to

land very close to them, or later human visitors destroy
them. Yet the Moon preserved no trace of any earlier visit
that could be found. To the would-be exo-archaeologists, it
was a profound disappointment.

Opinion changed, and the fashionable view questioned
strongly the 'assumption of mediocrity'. Earth became
again a special place, at least in some important respects.

Around the middle of the twentieth century experiments
showed that if energy, from more or less any source, is
applied to the chemical compounds that are believed to
have existed on the prebiotic Earth, amino acids are
produced. These are the building blocks from which proteins are constructed. Later it was discovered that among
those building blocks were some from which nucleic acids
are made, so that the basis of DNA, and the genetic code,
can appear in this way.

It seemed, then, that provided the chemical ingredients
are present in a fluid medium that allows them to react
together, and provided sufficient energy is supplied to
power the reactions, amino acids will appear. This first
step in evolution is probably quite common.

It is the second step, from complex organic molecules to
organized cells, that proved troublesome in those days, for
while it was perfectly possible to postulate mechanisms for
their formation, there were problems.

The processes could occur only in a fluid medium, but
the conservative scientific view was that they could occur
neither in the air nor in the surface waters of the seas, for
there they would have been exposed to ultraviolet radiation
that would have broken the molecular chains as fast as
they formed. They could have occurred in deeper water,
shielded from the ultraviolet, but there it would have been
the water pressure that broke the chains. The most likely
place for life to have begun was in tidal pools. Water

would have flowed in with the flood tide. Between tides much of the water would have evaporated, then more water would enter as the tide returned. In this way the ingredients needed to construct self-replicating nucleic acids would have been washed in, then concentrated, so that the bonds could have formed faster than ultraviolet radiation could have broken them, and finally the 'finished products' could have been washed back into the sea.

In other words, it was believed to be necessary to have tides, and for this the planet had to possess a satellite large enough to exert strong tidal forces. What was special about Earth was its location in the solar system, which exposes it to an amount of radiation that permits water to exist as a liquid on its surface, and its Moon.

Today orthodoxy has shifted its ground yet again, and Earth appears to us as mediocre as ever it did. In our view, most of the 'special' arguments are fallacious. They take an existing situation, calculate the probability of that situation appearing, and conclude that some extraordinary event, perhaps an outside intervention, must have forced processes in a particular direction. It is thinking backwards and the picture changes radically when you realize, for example, that the chance that the chemicals, the genetic material, the energy available on Earth and Mars would arrange themselves in such a way as to form me is exceedingly remote, but the chance that in fact I am me is one hundred per cent—it is certainty. No outside intervention is required to explain it. Each small step prepared the way for the next, and far from seeming extraordinary the process is almost inevitable.

We know that the power of ultraviolet radiation was both over-emphasized and, to some extent, misunderstood. People were frightened by this form of radiation. Where simple molecules are reacting, and reacting to form more complex molecules, the breakage of atomic bonds may be beneficial. It would speed the process substantially. Nor is ultraviolet especially harmful to simple life forms. Algae thrive in full-spectrum radiation. It seems to us most likely that life did begin in shallow water, but that it moved almost at once to moist sites on the land. There were clear advantages in living on the land, for in a large volume of water the greatest of all dangers is diffusion, of molecules simply being washed away and diluted to such an extent that reactions among them are slowed or stopped entirely.

Tidal pools might serve as places in which life could originate, but there are better places. There are sheltered, shallow creeks, for example, in which tidal movement is minimal, but where the evaporation of water draws in more sea water, and so concentrates the chemical ingredients of that water. Tides would be a positive disadvantage in such places, because they would bring water in only to remove it again a few hours later, and so the Moon becomes irrelevant.

If Earth is not special, we should expect life to commence on many planets that have water, an atmosphere, and sufficient energy from their star. It is reasonable to suppose that on some of those planets intelligent forms of life will evolve. That brings us back to Fermi's question, and the answer to it. Travel within a star system is fairly easy. Travel from one star system to another is not, and although it may not be impossible, it is impossible to imagine any way that a being with a lifespan comparable to that of any terran species could travel to another star within its own lifetime, far less go there and return. It is all too remote, and it may be as simple as that. No one can be bothered, can afford it, or dares, although we continue to listen for radio transmissions that will supply evidence of the existence of beings we may never meet face to face.

All of this may seem to be a digression, but such matters had to be considered very carefully when the human colonization of Mars was being planned. If the planet was to support life, then the planners had to know what that implied. Mars was known to have an atmosphere and water, and the fact that it has no satellite large enough to cause tides was unimportant.

The first people to visit Mars had a distinct advantage over the navigators who had explored Earth. Most of the martian surface had been mapped. They did not have to draw their own maps as they went, often not knowing what lay beyond the horizon.

Mars is rather more than half the size of Earth, with a radius at the equator of 3395 kilometres, compared with the 6378-kilometre equatorial radius of Earth. However, Mars is much less dense than Earth, so that gravity has only thirty-eight per cent of its terran value. Things weigh less, and this has allowed us to work at building our settlements with much less expenditure of energy than would have been needed on Earth. It also allows ships to leave our planet more easily than they can leave Earth. Our escape velocity is a mere 5 kilometres a second compared with the 11.2 kilometres per second needed to escape from Earth.

Attempts to map the planet began in the nineteenth century, as soon as astronomers had telescopes powerful enough to reveal surface features. Those early maps had to be abandoned, for they were too inaccurate to be of any use, but they established one tradition that has endured. In 1877, Schiaparelli, who also reported the existence of 'canali', drew a map in which prominent features were given Latin names, many of them derived from real or imaginary places on Earth. We have kept those names, and we have kept them in Latin, although more recently other place names have used other languages. The use of Latin was highly convenient. Being a 'dead' language, and the cultural property, as it were, of all mankind, its use avoided the nationalisms that so plagued the history of Earth, and the language was equally comprehensible—or incomprehensible—to everyone. Our children are accustomed to learning their latinized geography, and they can use maps printed by anyone. On Earth, even now, an English-speaking child might have trouble using a map drawn, say, in Czechoslovakia.

Mapping began in earnest as soon as really good photographs were available. The first pictures, from the Mariner missions of 1965 and 1969, showed only small portions of the martian surface, but Mariner 9, in 1971, revealed much more. In particular, it showed differences between the northern and southern hemispheres. The southern hemisphere is heavily cratered, the northern relatively free from craters. The southern hemisphere is mountainous, the northern more level and depressed. Much of it lies several kilometres lower than land to the south. Most martian settlements are in the northern hemisphere, in low latitudes not far from the equator, but I would not have you think the terrain is boring. True, there are broad plains formed from basaltic lava flows, and early explorers spent much of their time in a rolling, almost featureless landscape, but they knew from their maps that what they saw was not typical of the whole. Olympus Mons, at twenty-three kilometres high the largest volcano in the solar system, lies eighteen degrees north of the equator on the Syria Rise, and not far from three other giant volcanoes that form a straight line running south west to north east, dominating the Tharsis Montes range, which straddles the equator. To the west of the mountains we also have the Valles Marineris, a canyon system compared with which the Grand Canyon of Earth is the merest scratch on the landscape. Our canyon is about 4000 km long, and in places 700 km wide and 6 km deep. We do not lack tourist attractions!

Some of our volcanoes are still active, but they are different in certain important respects from volcanoes on Earth. For one thing, they are much older. This is because Mars is not constructed in the way Earth is constructed.

You might picture Earth as being formed from concentric layers. At the centre there is an inner core, surrounded by an outer core, both made from metal and rich in iron and nickel. The behaviour of these cores gives Earth its strong magnetic field and their density gives it its strong gravitational field. Beyond the outer core lies the mantle, a thick layer of rock, called magma, that is hot and held under such intense pressure that although extremely dense it behaves rather like a liquid. There are currents within it, and heat moves by convection. The outermost layer, the crust, is very thin, a film on the surface, and made from rock that has cooled and solidified. This magmatic rock has been weathered chemically, and by the action of frost, rain, and wind, changing it and breaking it into ever smaller fragments. The fragments have been deposited as sediments, compressed, heated from below, and deformed by movements of the crust.

The crust forms plates that ride on the magma below and move in relation to one another. Volcanoes appear over 'hot spots' in the magma, at places where the crust is weak, most commonly at plate margins. At some margins, one plate is disappearing beneath the adjacent plates, and where this happens volcanoes are demolished. At other margins new material is emerging from below so a plate is growing, and there volcanoes may be carried far from the 'hot spots' that caused them. Then they die, and the climate of Earth begins to erode them. These constant movements of the crustal plates limit the size to which any terran volcano may grow.

Mars is simpler. Its core is not rich in metals, and the planet has only a weak magnetic field. We do not navigate the martian countryside by means of magnetic compasses. The crust, on the other hand, is very thick indeed, and because it is so thick it is far too rigid to be broken into moving plates. Our surface stays where it is! This means that volcanoes can be neither destroyed nor carried away from their 'hot spots', so they continue to grow until so much material has been ejected that the vent is plugged, and that can take a long time.

Apart from volcanic craters, Mars has many impact craters, caused by lumps of rock and metal that have collided with the planet in the course of its history. Earth would have them, too, were its surface not so plastic, and were it not blanketed with vegetation and eroded by water. When Terrans were exploring the Moon, the ages of parts of the lunar surface were estimated by counting the craters, assuming a particular frequency for impacts and multiplying one number by the other. It is more complicated than it sounds, but when lunar rock samples were collected and could be dated radiometrically it proved fairly reliable. Attempts were made to estimate the age of our southern hemisphere in the same way, but on Mars the procedure is less reliable. Mars is still geologically active, and its surface is subject to wind erosion.

On Earth, crustal movements raise the mountains and cause rifts. We have rifts and deep depressions on Mars, but they are formed differently, by the vertical expansion and contraction of rocks within the crust. As rocks are heated, deep below the surface, so they expand and lift up the surface. As they cool, so the surface collapses.

In many places the land is broken, with high, sheer cliffs, and deep valleys. It was not too difficult to determine the height of cliffs and mountains from photographs. It is a matter of simple geometry and elementary trigonometry, no more. The angle through which the camera views its subject is known, because that is determined by the manufacturer of the camera and lens. A radar altimeter can measure accurately the distance between the camera and the ground. Thus the field of view can be known and the surface distances measured. Provided the local time of day is recorded when the photograph is taken, the angle of the Sun above the horizon can be calculated. With these items of information, shadows can be measured and the height of the objects causing them can be calculated. The martian maps began to acquire contour lines.

This worked well enough for features large enough to cast shadows the cameras could resolve, but there are many objects, down to the smallest pebble, whose shadows are far too small to be seen in a photograph taken from orbit. Yet even such tiny shadows provided information, for they darkened the surface on which they lay and allowed skilled interpreters to tell one kind of ground surface from another.

Other techniques were used. Our eyes are sensitive to electromagnetic radiation over only that small part of the spectrum we call 'visible light'. Light reflected from a surface can tell us much about that surface. The material from which the surface is made will absorb radiation in some wavelengths but not others. Mars appears red, for example, because its surface absorbs part of the visible light, but reflects in the wavebands we see as red. By using instruments that could 'see' over a much wider range of the spectrum than is accessible to the human eye, Earth-based explorers learned much more about the composition of the martian surface.

Mars had polar icecaps that advanced in winter and retreated in summer. They were visible to powerful telescopes on Earth. The Mariner 9 photographs revealed what looked very much like dry river beds, and there were many of them.

Clearly, there was no liquid water on the martian surface, but all attempts to devise alternative explanations for the river beds failed. They looked just like the wadis that can be seen in deserts on Earth, although there were certain important differences between the martian wadis and real, flowing, terran rivers.

If you use a map to follow a river on Earth from its source to its mouth you find that a main channel is joined by many tributaries, more or less of its own size, and that the junctions usually form large angles. Tributaries often join almost at right angles. On Mars, the main channel is more likely to be joined by tributaries much smaller than itself, and the junctions tend to form small angles. This is a pattern characteristic of channels that form rapidly, as they might on Earth following the heavy, flash rains that occur in deserts.

It was agreed that the martian channels are, indeed, dry river beds, and that at some time in the past water flowed across the surface. Some of the beds were obviously old. They entered craters where they vanished entirely, because the crater formed on top of them or because they were buried beneath wind-borne soil.

Water had flowed, but had there been rain? There may have been, but it is more likely that the water came from below than from above.

The first proposal, reasonably enough, was that in the distant past Mars had enjoyed a much warmer climate than the one observed in the twentieth century. The orbit of Mars is perturbed, slightly irregular, as is that of Earth, and the planet 'wobbles' a little, like a toy gyroscope that is slowing down. From time to time the exposure to solar radiation increases. The Sun itself is inconstant. From time to time it actually emits more radiation. Should the two phenomena coincide, so that Mars was more exposed just when the amount of solar radiation increased, then temperatures would rise, and rivers might flow.

The explanation was not very satisfactory, not least because since the formation of the solar system the Sun has grown steadily hotter. It is growing hotter still, with fluctuations over short periods, and will continue to do so until eventually it destroys its own planets. This means that the further back in time you look, the cooler the Sun was, and therefore the smaller the amount of solar radiation that planets received. There is no difficulty in finding explanations for cooler conditions in the martian past, but great difficulty in explaining warmer conditions if the Sun were the only source of that warmth.

It was the craters, and the discovery of large impacts on Earth, that provided a more plausible account. The Cretaceous Period of the history of Earth ended with the extinction of about three-quarters of the animal species alive at the time, caused by the impact of a planetismal. The object concerned, made from solid rock, was about ten or eleven kilometres in diameter and it struck Earth at a speed of about twenty kilometres per second (72,000 km/h, or about sixty times faster than the speed of sound). The damage to Earth was caused by the energy released by the impact.

On Mars such an impact would have released sufficient

heat to melt much of the ground ice. That would have caused the rivers to flow, suddenly and in torrents as the form of the river beds suggested, but not for long. Much of the water would have evaporated, and water vapour in the atmosphere tends to trap heat, because it is transparent to short-wave, incoming radiation, but opaque to long-wave radiation from the ground. Temperatures would have been kept above freezing for a time, but in the end they would have fallen again. Probably the first winter would have been enough to freeze the ground, water falling as rain or snow would have been trapped on the surface, the air would have become dry, and the climate would have returned to its former, stable condition.

Mars is not short of water. Our own ancestors would never have attempted colonization had the planet been truly arid. We will never have as much water as Earth, where three-quarters of the surface is covered by oceans, in places kilometres deep, but we are careful and we can manage with less.

You must remember that those first visits from un-manned craft explored a planet very different from Mars as it is today. It was extremely cold. The icecaps at the poles consisted partly of frozen carbon dioxide, with water ice beneath it. The hoar frost that appeared on the ground on autumn mornings was made of a mixture of carbon dioxide and water ice. Most of the water was locked below the surface, as permafrost.

The presence of permafrost was inferred from the photographs taken from space, which revealed areas of terrain that had collapsed as ground ice had melted and then frozen again in the past, a condition typical of permafrost regions on Earth. The permafrost contained enough water to make rivers flow and to fill our lakes, and in addition there is a very large reservoir of water beneath the martian pole.

The water was not all confined to the ground, even then. There were clouds in the martian skies, made from ice crystals, and with forms similar to the clouds in terran skies. Some of them were high, fifty to sixty kilometres above the ground. There was fog, too, as the ground frost evaporated and then condensed in the valleys early in the morning.

Though low, temperatures fluctuated widely. In particular, the great dust storms for which Mars was famous caused a marked warming of the atmosphere. As winds picked up dust, so the dust, composed largely of silica (sand), absorbed solar heat. This warmed the air, causing convection currents within it, and these in turn produced more winds to pick up more dust. The temperature at ground level could rise by as much as 25° C during a dust storm, and winds could reach hurricane force. The big storms tended to occur in spring, around latitude 40, as the southern icecap retreated rapidly owing to the sublimation of the carbon dioxide at its edges, so that the defrosted ground could be 80° colder than adjacent ground.

Even so, Mars was a cold place. The average temperature at the north pole was $-140°C$ ($-220°F$), and even at the equator it was only $-50°C$ ($-58°F$), although on a balmy summer afternoon in our temperate latitudes it could be as warm as $-30°C$ ($-22°F$). There were even reports of temperatures rising to well above the freezing temperature of water occasionally, to temperatures that would be thought warm even on Earth, although on Mars this did not allow the existence of liquid water. The atmospheric pressure was so low that any small, local body of water that melted evaporated at once, and its latent heat of evaporation chilled the ground on which it lay. Such slight warming would increase the concentration of atmospheric water vapour a little, but would not affect the permafrost.

The dust storms were started by winds of more than 100 km/h, but usually winds were light, at less than 30 km/h. The Viking landers recorded winds averaging about 9 km/h for the first three weeks they were there. Despite the storms, then, martian weather was generally not violent.

The air was thin, with less than one per cent of the sea-level atmospheric pressure on Earth, but pressure varied considerably. In spring, when solid carbon dioxide sublimed, air pressure increased, only to fall again as winter approached and the carbon dioxide froze.

Carbon dioxide was by far the most abundant gas in our atmosphere at that time. It accounted for ninety-five per cent of the air, nitrogen for about three per cent, argon for about one per cent, and oxygen, ozone, krypton, and xenon made up the balance. The modern atmosphere has a different composition, though it is still thin by terran standards. It changed first when the ground water melted, and water vapour entered the atmosphere. Then we added oxygen, obtained from the peculiar oxides that lay at the surface and made the planet red.

As the amount of oxygen increased, so did the ozone layer. You do not need much oxygen in an atmosphere to produce an ozone layer. If one-hundredth of the atmosphere is oxygen, an ozone layer will form, and that may be no more oxygen than can exist in chemical equilibrium in a mainly carbon dioxide atmosphere.

Ozone forms in the upper atmosphere as ultraviolet radiation supplies the energy to separate oxygen molecules into their single constituent atoms, and some of those atoms reform in threes (ozone) rather than twos (oxygen). The ozone and oxygen molecules constantly split and form again, and the layer of the atmosphere in which this occurs absorbs most of the incoming ultraviolet radiation.

The lack of an ozone layer was not a serious problem for the first colonists. They could protect themselves easily enough against the sunshine, and in any case the sunbathing that would have exposed them to possibly harmful doses had gone out of fashion. Their food crop plants were grown under cover, and the plants that grew in the open were not harmed.

Background radiation levels much higher than those encountered on Earth were a somewhat more serious threat.

Will they ever discover a cure for space-sickness? For reasons no one has been able to explain, I seem to be immune. I did not suffer in the least on the outward journey, and I have not suffered, so far, on the return. Almost everyone else has been laid low. All over the place passengers are strapped to bunks in considerable distress. Even crew numbers are depleted. For the time being I have the passenger areas of the ship almost to myself, and I am often alone at mealtimes.

The enforced isolation allows me time to reflect on the report I carry, and to collect my thoughts. I spend my hours trying to recall the history I was taught as a child, to draw beneath it, as it were, a line to mark the end of an epoch, the line I carry in the form of a small electronic data pack.

The first human visitors to Mars did not land. They were Russians, six of them. They made a few orbits, took many photographs, sent messages back to Earth, and went home. The Russians had been planning their Mars programme for many years, with orbiting stations that kept their cosmonauts for longer and longer periods in space, while on the ground they were devising ever larger rockets to carry materials into Earth orbit. They launched their manned probe in 1985, a few months before the first Foxe rocket left Earth to start the transformation. The Russian cosmonauts were the only humans ever to see Mars with their own eyes as it was before our history began, for we date our history from the arrival of the first consignment of CFC.

Obsolete rockets and canisters of CFC began to arrive at the Indian base in the autumn of 1984. The first launch took place a year later.

Before they could travel so far, the rockets had to be checked thoroughly, and then modified. They had been designed to carry a payload high into the atmosphere, but then to fall again. If they were to attain escape velocity they would need more power in relation to the load they carried. The solution was to use five rockets. Four of them were lashed together, the fifth, representing the payload, formed a second stage and carried the cargo.

It sounds crude, and it was, for the object originally had been only to rid Earth of the rockets. It was the meeting between Foxe and the scientists from the Oslo conference that altered the plans, and at a fairly late stage. There was neither time nor money for anything more elaborate.

Ties were made, with four curved sides, to which the four rockets of the first stage were fixed, so they were fastened together at the centre. Then metal hoops were passed around the outside and bonded to the rocket casings. The space enclosed by the rockets housed the wiring linking the rocket engines to the control mechanisms. The fifth rocket was fixed above what was now a four-engined rocket. It contained the control systems and cargo.

The four-engine rocket had enough fuel for the launch and for the first few minutes in space and as the fuel burned to a certain level a signal was sent automatically that ignited the second stage. After that any necessary course corrections could be made, during the short time that the second stage lasted before it ran out of fuel. Once they had fired, the engines could not be shut down, and would run until they exhausted their fuel, but the rate at which they burned could be controlled.

The companies that were so glad to be rid of their stocks of CFCs also supplied a substance known technically as dibromotetra-fluoroethane. It, too, was a chloro-fluoro-carbon compound, but one with a very high molecular weight. Once a solid fuel is ignited, it burns, and there is no way to regulate the rate at which it burns. However, small amounts of this very heavy CFC, injected into the jet of a burning solid-fuelled rocket engine, alter the molecular weight of the exhaust gases. The compound itself is destroyed instantly, of course, but the heavy atoms from which it was made are not. The local increase in weight causes a shock wave within the gases that are being expelled. With a small tank of the compound, and pipes from it to nozzles in each of the engines, it was possible to vector the thrust from each engine separately. It was quite subtle, for not only could a particular engine be made to 'push harder' than another, but the thrust of each engine could be directed this way or that. In this way the first stage rocket could be turned, for as long as its engines burned at all, and the second stage could be directed more precisely.

Although the weight of the payload had been reduced to enable the rockets to leave Earth, considerable power was available. The rockets must have looked terrible—no photographs of them have survived—but they worked. Each one was able to carry several tonnes of cargo, just securely enough to hold it together. Guidance, by means of radio

signals from the ground in response to information supplied by the on-site radar tracking station, was enough to
throw each rocket into an orbit that would carry it close
enough to Mars, and at a low enough speed, for it to fall to
the martian surface. That is all the rockets were meant to
do. They took off, they were set on their course, they
travelled, and they crashed. As they crashed, so each of
them released its CFC cargo. It was a low-cost enterprise!

The first rocket is believed to have reached Mars around
Christmas, 1985, and once the launch programme began
consignments were sent in batches every twenty-five days.
The launch procedure was simple and the site had many
strengthened pads, so that on a launch day it was possible
to despatch rockets at hourly intervals from dawn to dusk.

The Russians saw Mars as it had been, but the change was
rapid. The first humans to land saw a very different world.

Status Quo

After a few days, the passengers began to recover, one or two at a time. They returned, drifting and blundering, as they learned to manage their weightless bodies, pulling themselves along corridors gingerly, gripping the handrails tightly, and pausing for long deliberations whenever they met someone travelling in the opposite direction. They say you grow accustomed to weightlessness. As usual, they lie. How can you grow accustomed to a body that rebels chaotically against natural laws it has spent the whole of its life obeying?

In some ways it was worse for me than for the others. I had an idea what to expect, of course, a better idea than they had, but the very concept of 'weight' was becoming very confused to me. Naturally, at home, I weigh what a man of my age and height is supposed to weigh, around twenty-six kilograms. When I landed on Earth that proper weight was exaggerated arbitrarily to a crushing seventy kilograms. Hardly had I come to accept the elephantine load I had to drag with me wherever I went than it was time to return home, and the load was removed entirely. So what do I weigh? What does 'weight' mean? Should my legs be so unwise as to try to do their job and lift the mighty Earth body that was attached to them a short time ago, at best I would find myself floating helplessly in the air, probably out of reach of any handhold. At worst I could bruise myself painfully on what in a sane world I would have called the ceiling.

At least most of the public rooms in the ship have a top and a bottom—while you are inside them. The pictures on the walls and the furniture have to be one way up or the other, after all. In the cheapest, 'steerage' sleeping quarters, humans are stacked head to toe in rows across one wall, row above row like flies, if that is the way up you happen to be. There are little rails with curtains that can be drawn to give each of them some privacy, but there is little privacy while voyeurs can crawl about on the ceiling! The position of the door suggests that that particular wall might be regarded as the floor.

Maybe the crew gets used to all this. I suspect, though, that the people who really grow accustomed to weightlessness are those who sit at desks in travel agencies on one planet or the other, selling tickets.

The discomfort is temporary. That is a great consolation. The ship must be held still during docking and it remains in a constant attitude until its course has been programmed into the navigational computers, but then life improves. The gentle, sustained acceleration needed to move it across the solar system will soon supply us all with weight, and we will walk again on floors as, from our point of view, we rise vertically. Once the acceleration is completed and we achieve our cruising speed, the entire ship will begin to rotate, and the rotation will give us floors and weight with which to remain upon them.

For the time being it is sensible to wear a hard hat, supplied by the spaceline in several garish colours. At first, no one wore them for fear of looking foolish. Then I saw many people rubbing their heads. Then everyone was wearing them, everyone looked foolish together, but the heads were healing.

There is time to talk now. There are no windows as such, so there is no view of the stars outside, and we have several months full of hours to be passed. Certain of the walls are simulator screens. Just now they are blank, but soon they will be activated. At certain hours, I know from the outward trip, they will show films of a character sufficiently bland to cause no offence to the most delicate of passengers. At other times they will be linked to the outside cameras to show us the universe through which we are moving. We need a more enlivening source of entertainment. A diet of stars and ancient movies can pall rather quickly.

Sometimes a few others join our little discussions and I have the feeling that before much longer I will find myself delivering a series of lectures to quite large audiences. For the moment, we have a priest making his way to a small mission in Utopia, where he hopes to convert the community to a simple way of life based on the principles of Christian communism. He takes much interest in the moral debates that exercised Terrans in the twentieth century, while I am fascinated by the changes in outlook that occurred when finally the teachings of Christ, St Paul, and Marx all merged, towards the end of that century. There are a couple of chemical engineers, a horticulturist, and a drama director whose civilizing ambition it is to bring

Sophocles to Olympus. We have rather a good climate for outdoor performances of the Greek classics. Whether we can muster adequately large audiences is another matter.

They are young, all of them, and most of them adventurers in their own ways. They go seeking their new world inspired by the highest of romantic ideals. Perhaps our place names affect them. What they seek lies within themselves, of course. If they cannot find it there, a move to Mars will not reveal it for them magically. As Goethe is alleged to have remarked to the young student who poured out to the great man his enthusiasm for the new life he would find in America, 'If America exists, it is here.' When it comes to it, Mars is just a place, Martians merely people. We are not gods, despite the names on our maps. A volcanic mountain, even a very large one, is just a volcanic mountain. It is not the repository of some profound truth.

Today, though, my friends and I must spend a little time considering a matter that really is profound. We must discuss the nature of life itself, and the ways in which it manifests itself. It was in thinking around this problem that the first important comparisons were made between Earth and Mars, and humans knew for the first time with certainty that nothing lived or could live on or beneath the martian landscape. The planet was red and dead, and both facts were known long before the first human crossed the sky to see for himself.

We must get to grips with the concept of metastability. 'Stability' is a word with which you will be familiar. 'Meta' means 'behind' or 'beyond'. Metastability is a kind of stability that will endure unless something happens to nudge it. Imagine a gentle, smooth slope that goes down into a hollow, then rises over a rim, and continues to a lower level. If you roll a ball down the slope it will reach the hollow and, unless it has sufficient momentum to carry it over the rim, it will settle in the hollow and stay there until something comes along to push it the rest of the way. Metastability is like that. Chemically and physically, Mars used to be like that. It is not a final, chemical equilibrium, a homoeostatic condition in which a status quo is preserved that it is very difficult to disrupt, but more of a temporary state, and homoeorhetic, which means that many conditions within it remain constant while others change. Mars, after all, receives sunlight, and even in its primeval state this brought fluctuations in temperature, and a range of chemical reactions. Our success derived from our ability to 'nudge' the martian system, to produce a permanent change in the status quo by quite simple, gentle means.

Because we know a good deal about the internal structure of atoms, we know all the possible elements that can exist, we know how they behave in the company of other elements, and we know how much energy is needed to make them combine in this way or that. The laws of chemistry are fairly well understood. This means that if we know the composition of a particular substance, and if we know how much energy is being supplied to it, we can predict its chemical behaviour. The mathematics may be formidable if the substance is highly complex, and we may need to know the laws of physics as well, but at the very least we should be able to predict approximately the end, or equilibrium state, of the substance.

For our purposes, let us allow the 'substance' we are considering to be an entire planet, and to begin with let that planet be Mars, where a lifeless steady state was reached billions of years ago, and the tendency of the martian system to return to a state of chemical and physical stability was very strong. I mentioned earlier that whenever the surface and atmospheric temperature of Mars was raised, so that water flowed, the change was only temporary. In time stability returned. It was this great imperturbability of the system that preserved the status quo. (Obviously I must speak in the past tense, for Mars has changed.)

Someone may object, though, by suggesting that stability within a system is a condition to be desired. Would it not have been possible, for example, to bring about a situation in which a new metastable condition emerged, where temperatures were higher, water flowed, and the planet became habitable? The answer to that must be yes and no.

No one gave much thought to the possibility of providing a more congenial climate for Mars and just leaving it at that. Perhaps a new metastable level might have been achieved, though it is very unlikely, but the victory would have been Pyrrhic indeed, for the planet would have remained almost as inhospitable as before. You must understand that for life to be possible, there must be energy available for living organisms to exploit, an energy gradient if you like, and if energy is available in that way, then the system is not in equilibrium, is not metastable. It is not enough just to have energy. A planet might be very warm, but evenly so over all of its surface. Living things need to exploit the differences between the warmer and the cooler, between the higher and lower energy levels. Remember my analogy of the ball rolling down the incline, but held in the small hollow. It is not at the bottom of its slope, but until we push it there is no energy it can exploit to move further.

*   *   *

Yet some metastable states are easier to budge than others. You can see this clearly if you contrast Mars, as it was, and Venus, as it is and as in all probability it must remain.

It is understandable that throughout history humans have supposed that Venus might be a pleasant place to live. It is only very slightly smaller than Earth, so distances would be familiar and gravity would provide everything with a familiar weight. Being closer to the Sun than Earth it should be warmer, perhaps producing conditions rather like the terran tropics over most of the planetary surface. The planet is permanently shrouded in yellow-white cloud, so it seemed natural enough to suppose that cloud to be composed of water vapour, and Venus to be a rather humid place.

Alas, such delightful dreams were merely dreams. When instruments penetrated the clouds of Venus and measured conditions at the surface of the planet, a different picture emerged. Venus is much drier than Mars. The clouds are composed not of water vapour, but of particles of various chemicals, including sulphur compounds. The clouds are very transparent. They appear opaque to outsiders only because of the great depth of the atmosphere. Should you be so unfortunate as to find yourself standing at the surface, in the few moments remaining to you the land-scape could be seen clearly, but through a thin haze, and perhaps distorted visually by the great density of the air.

It would make little difference whether it were day or night. A day on Venus is equal to more than two hundred terran or martian days, but the dense atmosphere refracts enough light from the dark to the light side for the long night to be no more than a twilight. The planet rotates in the opposite direction from Earth or Mars, so that the Sun rises in the west and sets in the east.

The atmosphere consists mainly of carbon dioxide, and at the surface the pressure is about ninety times greater than the air pressure at sea level on Earth. That is one reason why you would not enjoy the view for long! The other reason is even more compelling. The surface temperature, by day and by night, is about 470°C (880°F).

While on Mars humans had to contrive a 'greenhouse' effect in order to warm the atmosphere, on Venus that effect occurred naturally, long ago, and nothing happened to halt it. It ran away, and probably it did so rapidly, because as the temperature rose the chemical reactions between atmospheric gases and rocks that controlled the composition of the atmosphere would have accelerated, so the entire planet soon reached its present thermal and chemical equilibrium. It is difficult to imagine how living things could survive at temperatures high enough to destroy complex organic molecules.

In order to cool the planet it would be necessary to rid it of much of its carbon dioxide. Photosynthesis is the process that would achieve this, but how are plants to survive there for long enough for them to make even the slightest difference? On Mars it was possible to darken the surface and so increase its absorption of solar heat. A lightening of the surface would have caused more heat to be reflected and so would have cooled the planet. Even this strategy cannot work on Venus. Direct sunshine does not penetrate to the surface, and at the top of the atmosphere the clouds are as pale, as reflective of heat, as they can be. We could not increase the albedo, the 'brightness', because already it is close to its maximum.

Even if all this were to be achieved in some way, the atmosphere remains dense. How could the pressure be reduced? Could gases be lost into space? The way to lose gases from a planetary atmosphere is to heat the atmosphere, so imparting sufficient energy to its constituent molecules to permit the lighter of them to achieve escape velocity. The atmosphere of Venus is hot now. Gases that might be lost in this way departed long ago—and the hydrogen and hydroxyl (hydrogen-oxygen) molecules derived from the dissociation of water were among them. Even if the temperature could be reduced, and the pressure could be reduced, Venus remains immensely arid. Water probably would have to be imported, for terranmartian life requires water.

Venus, then, presents a much bigger problem than Mars ever did and so far we have not the merest glimmering of an answer to it. It may be solved, of course, one day, but that day is distant.

If all you want to do is inhabit a planet, it is not too difficult. The Moon is habitable, after all, and became so the moment humans were able to travel to it. People have been living there for years and conditions for them are not unpleasant. You need only two things to make the Moon, or primitive Mars for that matter, quite tolerable. The first thing you must do is to provide all the humans with protective suits. These suits must be hermetically sealed, must carry an air supply, must allow the body temperature of the wearer to be held constant, and if they are to be worn for any length of time they must allow food to be introduced to them and bodily wastes to be collected and stored for disposal later. Thus clad, people can go nearly anywhere. They have 'walked' in space, so you can see you do not even need a planet. While wearing their suits, the humans can pass the time building surface or underground establishments within which they will be able to remove their suits. The buildings must be hermetically sealed, provided with air, heated or cooled as appropriate, and within them the humans will be able to keep such terran plants and animals as they need to supply them with food, fibre, or other agricultural products. In other words, you can live almost anywhere, provided you can supply yourself with a little capsule that contains as much of Earth as your survival and comfort require. The larger environment, beyond the walls of your capsule, can be as metastable as you like, but you may never enter it unprotected and hope to return alive. On the Moon, death comes swiftly to the naked human.

That is the part of my answer I said would be 'yes'. This is the sense in which a metastable environment may be habitable. My 'no' is really an answer to a slightly different question, so I will rephrase it.

Is it possible to conceive of a metastable environment that can support life while remaining metastable? No, it is not. What is more, my denial of that possibility embraces not only terran life, but all life that conforms to the broad definition of a living organism that is applied on Earth and Mars—and that definition is pretty comprehensive.

Without risking statements about anything so abstract as life itself, we can say something about organisms we all agree are alive. They obtain chemical substances from outside themselves. They use these substances to repair and renew their own tissue, and for growth. This requires energy, and they obtain energy either chemically, and also from substances they obtain from outside themselves, or directly from sunlight. In using all these substances they produce wastes, of which they rid themselves. They reproduce themselves, and, finally, they die and the substances of which they are composed become the raw materials for other organisms to continue the cycle.

You may allow that this describes terran and martian life forms, but are these the only life forms we can imagine? Does everything have to work in this way? Well, we might imagine a life form that used a different kind of chemistry, not making much use of carbon say, but that changes little. Substances are still being taken in, processed, excreted, even though the substances themselves may be different. There might be organisms that found some novel source of energy to exploit. On the deep ocean floors of Earth, far below the deepest limit to which sunlight can penetrate, there are organisms living very satisfactorily by making use of a local energy gradient. Chemical substances entering the ocean from the magma at certain tectonic plate margins are collected and allowed to react together to release energy. These animals are startlingly different from surface organisms, but they obey the rules, as it were. They take something and make it into something else—themselves.

What, though, of those life forms that consist of pure energy, having abandoned their physical bodies, the beings so beloved of certain writers of science fiction? Such eminences engage in no chemistry. If they reproduce themselves, which is uncertain, they do so by novel means, but even they do not break all the rules. They 'do things'. If they did not 'do things', if they were totally and perpetually inert, why should we consider them to be alive at all? If they 'do things' they need energy, which they must obtain from a source. Even if all they do is to exist, as pure energy concentrated locally—some kind of 'ball' of energy—they must expend energy to sustain themselves. If they could exist without adding in any way to the energy they had at the start, they would be violating the second law of thermodynamics, and that would be quite a trick! The second law states that energy will distribute itself evenly throughout a closed system. It is why hot water cools when left to stand in cooler air. If the water is to be kept hot, or living organisms are to retain energy, more energy must be applied. The water will need a heater, the living organism a form of chemical energy—food.

I realize we have a difficulty here. If there were, or could be, life forms that existed in a way so alien to us that they did not engage at all in the physics and chemistry of their environment, then there is little we can say about them except that to us they are utterly incomprehensible. We cannot imagine them. Probably we would not recognize them as living even if we were to encounter them. There is, after all, a limit to our understanding. Our minds have bounds. Sir Fred Hoyle, an eminent British astronomer of the twentieth century, once said that there are fish swimming in terran seas that can have no possible understanding of the human societies existing in the towns but a few metres from their habitat. Fish are intelligent, say by the standard of a worm. Perhaps there are beings as far beyond our comprehension as humans are beyond the comprehension of fish. There was a film once, called *2001*, in which humans encountered aliens so alien that they, and the audience, had no idea what was happening. It made the point.

The fact remains that if we accept as 'living' only those organisms that have some kind of a relationship with their environment, a consequence emerges. No matter how they live, no matter what chemistry they employ or source of energy they exploit, in their relationship with their environment they modify that environment. They must, because they use it, depend on it, and are part of it. Terrans are made from the stuff of Earth. I am made from the stuff of Mars, but by means of a chemistry imported from Earth. When I breathe I make a tiny change in the chemical composition of the air around me. The oxygen I inhale is used to warm my body, and the warmth of my body warms the air in contact with it. I have a very special relationship with the cabbages. In collaboration with them, I can turn cabbage into me, and my bodily wastes back into cabbage again. In doing so I modify the chemistry of the soil and water.

You and I do this constantly. In the literal, biological sense it is what we mean by 'living', but we are not alone. Indeed, we are not even very important. What matters in a particular environment is the total mass of all the living things that are engaged in this kind of work. In this spaceship, we humans account for a fairly large proportion of that mass, but on a real planet, a living one, we are insignificant. On Earth, for example, the weight of the organisms living in the top few centimetres of a field of grass is much greater than the weight of the cows feeding on that grass. You might stock five cows, weighing say 2.5 tonnes, on one hectare of very good pasture. Depending on the soil, the population in the top few centimetres may weigh between 11 and about 22 tonnes per hectare, or around 1.6 kg per cubic metre, and of that total, more than 1.4 kg consists of nothing but bacteria, fungi, and protozoa. On Earth, the total weight of all the organisms that are too small to be seen by the unaided human eye exceeds by a huge margin the weight of those you can see.

When you add together the effect on the environment of each of these tiny organisms it amounts to a major alteration in the chemistry of the entire planet. It is this alteration that allows us to distinguish between a planet that supports life and one that does not, and we can make the distinction even if the living organisms themselves are hidden from our sight. It is by their deeds that we know them!

If you look at it from another point of view, in their various ways the organisms are exploiting energy—almost all of it from the Sun—and in doing so they are passing it about among themselves in the form of chemical energy. They can do this because they have available to them chemical compounds in which energy can be stored temporarily. Some of these have long names, but others are very familiar. You can burn a log of wood. When you do so, the carbon that makes up a large part of the bulk of the wood is oxidized, and this reaction releases energy as heat and light. The wood was storing energy. Because it was storing it, the wood was not in chemical equilibrium. Equilibrium would be reached when all the carbon had been oxidized, and there was nothing left to burn.

Let us take the case of Earth, for it was by studying Earth as it was and as it might be that scientists were able to make the comparisons with other planets which told them that Mars, primitive Mars, was lifeless.

You will recall that for centuries there had been speculation about life on Mars, and that when the unmanned exploration of the planet began it was the question of life that aroused the greatest popular interest. Since it was the popular interest that persuaded governments to provide funds, there was a strong incentive to find answers to that old question. Today we might argue along such lines as:

'We do not know whether there is life on Mars, but we do know that there is life on Earth. We cannot say what "life" itself is, but perhaps there is some characteristic of Earth which distinguishes it as life-supporting. Let us examine Earth, therefore, as though we were seeing it for the first time and from afar, to discover whether we can detect life on our own planet and if so, how.'

In those days the only thing to do was to go and look.

We have been travelling for some days, and were we able to look through windows on the sky outside our ship we would find that Earth appears as no more than a rather large, very bright 'star'. Let us suppose we have aboard such instruments as we would need, and see whether life can be discerned on Earth.

The first thing to do is to calculate the amount of energy the planet receives from the Sun. This is not too difficult, for we know a good deal about the Sun, we can measure its output, and we know the orbit of Earth and hence the distance between planet and star for any day of the terran year. This calculation will give us a number representing the energy that is available to power chemical reactions.

The next stage involves using a technique, and equipment, that were used as long ago as 1967, to examine Venus.

We will examine Earth using an infra-red telescope, combined with a device (a Multiplex Interferometric Fourier Spectrometer, but the name is not important) that allows us to analyze in great detail the chemical composition of the atmosphere by means of a study of the spectrum into which light from it can be broken. We can also measure the density of the atmosphere and its total volume. That will allow us to calculate the surface atmospheric pressure.

We would find that the atmosphere of Earth is composed of 78 per cent nitrogen, 21 per cent oxygen, 1 per cent argon, and about 1.5 per cent methane, with a very small amount of carbon dioxide, a variable amount of water vapour, and traces of a number of other gases and solid particles.

We can measure the atmospheric temperature, and by the amount of water vapour and the way it varies, we can conclude that much of the surface of the planet lies beneath great oceans.

Given the amount of solar energy the planet receives, methane and oxygen cannot exist together. They will react, at a known rate, to produce carbon dioxide and water. So we can work out the rate at which these two gases are depleting one another. When we check our calculations, however, we discover that far from depleting, the amounts of the two gases remain remarkably constant. This must mean that both are being replenished. We know water is present, and so it is possible that sunlight may be dissociating hydrogen and oxygen in the upper air, but that reaction would need to produce one hundred times more oxygen than we calculate it does if it is to make good the loss through the reaction with methane. Anyway, how is the methane being produced? Perhaps it is belching forth from volcanoes, but we know of no volcanoes that are in a state of permanent eruption, which they will need to be if they are to keep the methane in the atmosphere. If the methane is being produced from the only raw materials we can find, carbon dioxide and water, that reaction must proceed through at least four stages in each of which the product is more easily oxidized than the methane itself. It is unlikely that any one of the four steps would take place from purely chemical causes, and that all four would do so stretches credulity too far. It just would not work. There are other carbon compounds, too, some of which are much more complex than methane and whose formation in the presence of oxygen is virtually impossible.

This is curious, but there is more to come. The existence of free oxygen is curious, and the amount of oxygen is still more so. There is twenty-one per cent. Now oxygen is dangerous stuff. It reacts very readily with a wide range of elements to form oxides that are quite stable. Therefore, there should be very little oxygen in an atmosphere. It should all be 'locked up' in those stable oxides. If there were a little more oxygen, say twenty-five per cent, that is what would happen. All the carbon would be oxidized very quickly. All carbon-containing substances would burn vigorously and Earth would become a stable cinder. It is not a stable cinder, but with an amount of oxygen close to that critical level we had better take fire extinguishers with us if we plan to land, for fires must be very common and sometimes extensive. If there were much less oxygen, on the other hand, less than twelve per cent, nothing would burn at all. The amount present, the twenty-one per cent, is enough to make fires common, but not enough to allow them to get completely beyond control.

Oxygen is also very poisonous, because it is so reactive. It is not easy to live in an oxygen-rich environment. Even humans have not adapted to it fully. The body of a mammal produces enzymes to prevent damage from oxygen, but even so, in the course of a full human lifespan, the quantity of oxygen to which the body is exposed is just about enough to cause cancer.

It does look, though, as if Earth supports some form of life based on carbon and using oxygen and stored, oxidizable, carbon compounds, to provide energy. The amount of oxygen available could supply sufficient energy for quite vigorous activity. If there are living things on Earth, some of them are likely to run fast, and some may even fly.

All of this suggests that Earth may support life, but only by saying that it provides conditions living things might find congenial. The more persuasive argument rests on a comparison between the conditions we can see and measure, and the conditions the laws of chemistry and physics would predict. If we allow purely chemical reactions to run to their conclusions in the atmosphere we have analyzed, and if we take account of the amount of carbon at the surface of the planet, which we can also detect as well as infer, we can draw up an inventory of what the atmosphere should contain. It should consist of about ninety-nine per cent carbon dioxide, and one per cent argon, with no nitrogen and perhaps just a trace of oxygen.

Under these circumstances, probably the oceans would disappear, as the 'greenhouse effect' caused by the increase in atmospheric carbon dioxide raised the surface temperature to, and then beyond, the boiling point of water. In that case the air would contain much water vapour, forming steam clouds at high altitudes. If the oceans were still there, they would be ten times more salt than in fact they are, and they would contain appreciable amounts of sodium nitrate.

We could have made a mistake in our calculations, but the important point is that any calculation of what a metastable Earth might be like leads us to a model that is vastly different from the Earth we observe. Since the planet is in nothing like chemical equilibrium, and is being maintained in an unstable state, we may conclude that it supports life. Something, some process or set of processes, is exploiting solar energy and using it to store chemical energy. It has turned the entire surface of the planet into a vast chemical factory and, having achieved that, it is able to keep its factory running by ensuring that conditions remain congenial for its operations. This amounts to something rather close to a definition of the behaviour of living organisms.

If we allow that this is what is happening on Earth, we can begin to discern more of the fine detail. The methane, for example, acquires a purpose. It is oxidized, and so removes oxygen from the air. If it did not do so, then little by little the amount of oxygen would increase, until the great conflagration took place that would make the continuation of life impossible. If you want to worry about environmental pollution, consider the consequences of oxygen pollution! Where is the methane produced? By micro-organisms, of course, and some of them live in the guts of large animals, such as humans.

This regulation of the amount of oxygen in the air has another effect, no less beneficial to the organisms that live on Earth. We know, because we are well informed about the life stories of stars such as the Sun, that for about the last four billion years the Sun has been growing steadily warmer. We can calculate that if Earth were like primitive Mars or present day Venus, allowing for its location in the solar system, it should have a surface temperature of around 300°C (572°F). It does not, and what is more, the surface temperature on Earth has changed hardly at all during all this time that the Sun has been growing hotter. True, there have been major glaciations and periods when the planet enjoyed a more benign climate, but because of the strong positive feedback in the mechanisms that cause the expansion and contraction of ice sheets, changes of only a few degrees in average temperatures are needed to initiate or end a glaciation.

This thermal regulation is achieved by the manipulation of atmospheric carbon dioxide. If the climate becomes warmer, plants grow more abundantly and life generally accelerates. This uses more carbon, obtained in the first instance from carbon dioxide in the air, and so reduces the proportion of carbon dioxide in relation to oxygen. Carbon dioxide is a 'greenhouse' gas, which traps heat beneath it, and so reducing the amount of carbon dioxide in the air allows more heat to escape.

The plants that abstract carbon dioxide from the air move much of the carbon to below the soil surface. A plant may have a root system that is more extensive, and accounts for a larger proportion of the total mass of the plant, than the stems, leaves and flowers that appear above ground. As organisms die, most of them are decomposed

in the soil or, in the case of aquatic habitats, in the
sediments. This releases carbon dioxide into the soil, and
over the entire Earth, the soil contains several times more
carbon dioxide than the whole of the atmosphere. When
conditions are right, this carbon dioxide can combine with
minerals consisting of calcium silicates to form the stable,
insoluble material we know as limestone. All the carbon
on Earth was released originally by volcanic eruption, and
volcanoes continue to emit carbon dioxide. The only final
'resting place' for carbon is in limestone. So there is a
cycle that begins with volcanic eruptions and ends with the
formation of limestone.

When the climate cools, the growth of living things is
slowed, but the decomposition of dead organisms continues. The release of carbon dioxide in the soil continues,
but abstraction of the gas from the air is reduced. The
oxidation of carbon in soil consumes oxygen, and the
surplus carbon dioxide is released into the air, where it
forms a blanket trapping heat, and so reduces the heat loss.

As the Sun has grown hotter, it has been necessary to
cool the Earth rather than warm it, and so over the millennia
carbon has been locked away safely in limestone. Probably
you could calculate the amounts of limestone formed
during different periods of Earth history and relate its
formation to the increase in solar output. In times of rapid
climate change, however, vast amounts of carbon may be
buried rapidly, unoxidized, so removing them from circulation. On Earth this buried carbon takes the form of coal,
peat, oil, and natural gas.

The living things of the living planet work together,
components of a huge system to maintain themselves and
their planet. If you could look really closely, there are
countless ways in which this collaboration allows organisms to help one another. Take iodine, for example,

without which humans become very unhealthy. The iodine comes from the sea and eventually is washed back into the sea. How does it move from the sea to the land? It is concentrated by certain seaweeds which collect it from the water, and from the seaweeds it is released into the air. Take common salt. There is some mechanism, which even now we do not really understand, by which salt is moved from the sea back to the land. As I said a little while ago, if Earth were dead, the sea would be ten times saltier than it is. If it were twice as salty as it is now—and as it has been for millions and millions of years—then almost all the organisms living in it would die.

From our distant vantage point there are two reasons why it is the atmosphere of a planet we should study rather than the liquid or solid planetary surface. In the first place our sensors would need to probe the atmosphere if they were to examine anything beneath it, and this complicates the technique. It is not the main reason, though. The mass of the atmosphere of Earth-like planets is a tiny fraction of the mass of the upper layers of the surface—soil or water—that support life. All the elements upon which living organisms depend move in biogeochemical cycles, which carry them between the organisms themselves and the non-living parts of their environment, and the great majority of them must pass through the air at some stage in the cycle. Since the mass of the atmosphere is relatively small, the turnover rate is greater there, elements remain for shorter periods, and consequently the activity is easier to detect.

If the sensors on our ship were really acute, there is one more thing we might discover about the distant Earth. If we could detect the presence in the atmosphere of substances that are there in tiny amounts, then among them we might find some substances that would strike us as strange. We might find organochlorines, for example, or even those chloro-fluoro-carbon compounds that played such an important part in our history. These are compounds that cannot be formed naturally under the temperatures and pressures we know prevail on Earth. They must have been made deliberately and then released. A living organism that is able to generate the very high temperatures and pressures needed to manufacture such chemical compounds and that then releases them into its own environment must be intelligent. We could detect the presence of intelligent beings on the planet, long before our best telescopes could reveal roads, canals, or even the lights of the cities at night.

The more perceptive terran scientists concluded, then, that a planet which supports life will cause so great a disturbance to its chemistry as to move the planet far from equilibrium. The difference between such a planet and one like primitive Mars can be detected easily and swiftly by anyone with the instruments necessary to analyze the atmosphere of the planet, and a knowledge of the laws of chemistry sufficient to predict, more or less, the equilibrium those atmospheric constituents should reach. It was as simple, or as difficult, as that. The only exception might be found on a planet where life was just appearing for the first time, so that the changes it made were barely perceptible. Mars was not such a world.

So you see that from long before anyone actually set foot on martian soil the ingredients were provided for a big debate about ways in which Mars might be made habitable. Until then, traditionally you might say, two views had prevailed. One held that Mars supported life anyway, so that humans need do little more than make minor adjustments, perhaps by finding ways to make its indigenous flora and fauna edible. Colonization would be mainly a matter of 'taming' or 'domesticating' the planet, much as the Americas and Australia had been domesticated. That view had to be abandoned the moment it was confirmed that Mars was lifeless.

The alternative view held that Mars was inherently uninhabitable and that therefore humans would need to live inside protective enclosures, little models of Earth that they carried with them. On that kind of Mars people would live as they live on the Moon, or as they live in space-ships. Eventually, perhaps, these enclosures would be expanded to occupy more and more of the space available for them, until Mars supported a population density comparable to that of Earth, and all contained within a simulated Earth.

People did live in that way for quite a long time, of course, but it was not the final objective. The newer view of the relationship between living organisms and their environment suggested a quite different way to approach the problem. If the planet could be 'seeded' in some way, if the conditions needed for life to begin could be supplied and then living organisms could be introduced, in time Mars would turn itself into a place where humans could live in the open, without special protection. With a little help, and it might turn out to be a very little help indeed, Mars might be transformed into a living planet.

Unfortunately, the planning was in the hands of humans, and this presented a problem. Individual humans do not live for very long and they do like to see some results from their work. They can be altruistic on occasion, providing for future generations benefits they will not be able to enjoy themselves, but this is rare, and the transformation of Mars by this means asked too much of them. You see, it took perhaps a thousand million years from the time it was formed until Earth began to support life of any kind. From that point, which we might take as the starting point for Mars, it took a further three thousand million years or so before the earliest living organisms had prepared the planet for the larger plants and animals which followed. It was an awfully long time to ask people to wait. They were not so unrealistic as to wish to embark for the planet tomorrow, or even next week, but in a year or two perhaps. Thousands of millions of years was too long a time. The idea was fantastic. Can you imagine how anyone could persuade investors to pour capital into a project that would realize tremendous profits, thousands of per cent return, provided only that the investor wait a few thousand million years? There are more attractive schemes on offer. Even planting oak trees and yew trees will do better than that. They will not mature in the lifetime of the person who plants them, but at least you can expect that there will still be humans around when they do mature. Some of those humans might even be related to you. If you go back into Earth history just one thousand million years you will find no humans, or large animals of any kind come to that. Why should you suppose there will still be humans a thousand million years in the future?

Someone had to think of something better, something that would work more quickly. So that is what someone did.

The new idea was to combine the reformation of Mars with old-style colonies living in bunker-like protection, so that the new inhabitants could superintend the transformation, accelerate it dramatically, and, because they were on the spot, take advantage of new ideas and opportunities as these presented themselves. It allowed for the planet to be changed, as it were naturally, but also placed humans, and the non-humans they took with them, to work at the problem directly—and we must not underrate the probability that big advances will be made as a result of the activities of non-humans. I doubt whether anyone thought of the scheme in such cruel terms, but it amounted to recruiting human volunteers and slave non-humans, transporting them to Mars, putting them in an enclosed, almost penal setting, and leaving it largely to them to work out how to escape. If that was the stick, the carrot of the enterprise was contained in all the dreams of new lives, all the hopes of wealth, freedom, adventure, that were locked in the minds and hearts of those first pioneers.

What happened next? What happened next was what is happening to all of us now. People actually began to move themselves to Mars. It was not then as it is now. You (and I) may find this spaceship cramped, uncomfortable, the journey boring—except when we are talking, of course— but our discomfort, the dangers to which even now we may be exposed should the Sun be going through one of its periodical bursts of hyperactivity and bathing us in its radiation, are as nothing compared to what was endured by the early travellers. When next we meet I shall tell you travellers' tales to chill your hearts.

The first travellers, though, were not human at all. They were the Foxe rockets, loaded with CFCs. A reasonably large proportion of them reached their destination. Some of those that missed left the solar system altogether and are swinging their way through interstellar space to this day, all ready to do their work on any planet they may encounter— perhaps to the benefit of alien life, for all we know.

Those that arrived safely (if you can call it that) crashed on the surface. People still find bits of them now and then and our museums have so many relics that if you find what may be Foxe rocket, you can keep it. We have enough!

When they crashed they released their cargo, whose fragile containers shattered on impact. The CFCs entered the atmosphere, were dispersed over the planet, and at once they began to do what those Oslo chemists had known they would do.

CFCs are 'super-greenhouse' gases. Like carbon dioxide and water vapour, they are transparent to short-wave radiation and fairly opaque to long-wave radiation, but they trap heat about a thousand times more effectively than does carbon dioxide. Turn that statement around and you will see you need one thousand times more carbon dioxide than CFC to produce the same warming effect.

I have mentioned these 'greenhouse' gases several times, but I do not think I ever explained just how they work. The Sun emits short-wave radiation which reaches the ground surface. The surface is warmed by it and begins to radiate its warmth back into the sky, but as long-wave radiation. This is trapped by the molecules of 'greenhouse' gases, which are warmed, and which then radiate their own warmth, a proportion of it downward, back towards the surface. Most dust particles do the same thing. Such of their radiation as is emitted upward is likely to encounter more 'greenhouse' molecules or particles, and they, too, are warmed. So the process continues, and you can see that if there are enough suitable molecules or particles in the air it can make the escape of long-wave radiation difficult. The air is warmed, the ground is warmed, and once the process begins it will continue, all by itself.

The world to which the early pioneers travelled was not immediately habitable. It was not yet Earth-like, but neither was it the icy desert that primitive Mars had been. It had warmed, so you could walk on the surface in ordinary terran clothes, even though you did need breathing apparatus, and there was water.

Even terran history lessons should have impressed on you the date of the first voyage by humans to Mars. The first of my ancestors arrived on the planet on May 17, 1997, by Earth reckoning, and possibly to the surprise of many students of space exploration, neither he, nor any member of the crew of his spaceship, nor any member of the ground staff or administration who sent him, belonged to a military organization. The armies, navies and air forces of Earth visited Mars, but they never did secure a foothold there. On Mars you will meet no soldiers and, as I expect you know, the writ of the rulers of Earth does not run on Mars, and never did.

Travels in
Space and Time

If you ask me, no one on this ship has the remotest idea what time it is. People arrive late. When they do not arrive late, they arrive early. Meals are times of chaos rather than the solemn ceremonies for the practice of a gentle art that some ancient Zen master once suggested they should be.

The trouble, obviously, is that all clocks and watches were left behind on Earth, except for mine. I left mine on Mars. Time is quite different on the two planets.

All the same, when I do meet with my travelling companions I am able to amuse them. They found it strange that a man of my configuration should weigh twenty-six kilograms, or fifty-seven pounds as some people prefer to call it. It is a perfectly proper weight, but they consider it, even now, as 'a mere' twenty-six kilograms. I built on that by telling them my age. I pointed out to them that my hair is greying (in case they had not noticed) and admitted that my waistline is somewhat larger than once it was. When people look at me they regard me as typically middle-aged, and so I am. I am a grandfather, too, and I am twenty-six years old. This announcement is guaranteed to bring roars of laughter from those who hear it for the first time. So I go on to say that at my next birthday I will be twenty-seven, and that by the time I am in my middle-thirties I hope to be a great-grandfather.

They deal with all this in the instruction they issue to would-be migrants back on Earth, but people do not take in the information fed to them. They are much more interested in the wealth they hope to acquire, the living standards to which they aspire, to pay much attention to bits of apparently trivial arithmetic. At this point, therefore, some of my audience begins to look nervous. Does Mars emit some strange gas, some Dread Ray, that accelerates the ageing process? Are we all embarked with one-way tickets to early graves? It is only adults who can be tormented in this way. The children are far too sensible and far too well informed. I do not even try it with them, although I do discuss the way things work.

There is nothing magically sinister about Mars. It is just that it takes almost two Earth years for the planet to complete its orbit of the Sun, and it takes a little more than twenty-four hours for it to complete a rotation on its own axis. It is arithmetic, not magic! Were I a Terran, I would be fifty years old, not twenty-six. We are no younger, we just get fewer birthdays. That, incidentally, is why we make such a big fuss about birthdays. They are the occasion for celebrations much more elaborate, and perhaps even rowdier, than those on Earth.

It is the difference in the length of the martian and terran days that makes watches and clocks what you might call planet-specific. It is neither difficult nor expensive to make watches that record times and dates for both planets, so you can read one or the other at the flick of a switch, but for most people these are toys, and have no more than a novelty value, rather like a watch that could tell you the time in different parts of the planet. In the real world—either world—you do not need all this information. Of course, you can buy martian timepieces on the ship, but hardly anyone does. True or not, the belief is that everything on a ship costs nearly twice what it would cost on the ground. It is an exaggeration, but ship prices certainly are high. The predictable result is that the three terran and one martian spaceports are crowded with people doing a brisk trade in the sale of watches from new arrivals and the purchase of watches from departing passengers. I gather it is quite profitable, if the traders will allow you a pitch on which to work. It is these traders, and the large printed signs in the passenger lounges, that remind people to remove their watches; even they make no mention of the reason for it. Perhaps passengers imagine a watch might interfere with the operation of electronic equipment on the ship?

The time we spend in space between the two worlds is used to acclimatize travellers to the very different conditions they will meet on arrival, and time adjustment is the first stage in that process. It begins as soon as you are on board. Ships travelling from Earth to Mars use martian time and dates throughout; those going from Mars to Earth use the terran system of reckoning. Everyone must depend on the time and date displays that the ship provides, but although many are provided they are not in every room. People cannot find one when they need it, or they forget to look, or they rely on their own internal sense of the passage of time, and they arrive late or early and cross.

It is time now to prepare for the discussion we will have later in the day.

Mars has a year of 687 days. To be more precise, it is 686.9804 days, which means we have 'leap years' every so often to set the calendar straight. Our 'leap years' are not like those on Earth, though, which is why I use the expression in quotation marks. On Mars leap years come every fifty-one years, and we lose one day.

Because our year is so long, compared to a terran year, our seasons, too, are long. Most Martians live in the northern hemisphere, where the winter lasts for 160 days, spring for 199 days, summer for 182 days, and fall for 146 days. These are average times, of course. In the southern hemisphere the seasons are a little different. Winter lasts 182 days, spring 146 days, summer 160 days, and fall 199 days. To some people this may make the seasons sound tedious, although I dare say there are those who would welcome a spring that lasts some 28 weeks. Unfortunately, the martian climate is even more tedious than it sounds. We all live close to the equator, where there is hardly any change in the weather as one season gives way to the next. If the plants notice they have said nothing. They just carry on growing.

It is because our martian day is rather longer than a terran day that the timepieces have to be changed. If you take a terran watch to Mars it will gain crazily, and if you take a martian watch to Earth it will lose. By the terran reckoning, our day lasts for 24 hours, 39 minutes, and 35 seconds. Of course, we have a '24-hour day' just as people on Earth, but each of our hours, minutes and seconds lasts a little longer than their terran equivalents.

It is a convenient coincidence that a martian day is so similar to a terran day, for it is the disturbance in the cycle of day and night and the consequent disruption of the diurnal rhythms which regulate many processes in the body that causes 'jet lag'. This could have been a real problem. On Earth, air-crews who fly regularly on east-west routes suffer seriously and their lives are shortened, but for ordinary passengers the effects wear off soon after the journey ends. If you moved to another planet where the day length was markedly different the 'journey' would not end at all, and the body might or might not be able to adjust satisfactorily to the change. On Venus, for example, one day is as long as 243 terran days.

As it is, there is some small effect, and precautions can be taken to minimize it. Passengers are advised not to eat for twenty-four hours before embarking, but to drink coffee, tea, and other beverages containing caffeine. For some reason that I do not understand, this helps.

I had better say something about the martian calendar. It is much simpler than the calendar used on Earth, but even so it seems to cause difficulties, and in truth confusion is possible.

We do not count months. On Earth these are based, clumsily, on the orbit of the Moon. On Mars we have no such conspicuous Moon. Indeed, we have two tiny moons that look about the size Venus looks when seen from Earth. A division of the year into months would force us to choose one in preference to the other, and that would cause endless wrangling among the Phobos and Deimos factions that would spring up instantly. Even then it would not be easy. Phobos orbits Mars three times each day, and Deimos takes rather more than one day to make a single orbit. Martian months would be rather different from terran months! Perhaps we could use both and try to devise a double-month system. I cannot begin to imagine what that would be like. We do divide our days into batches of

seven, making weeks, and we give the days their terran
names. A seven-day week is convenient. We write the date
as the number of the day in the week (one to seven),
followed by the number of the week in the year, followed
by the last two digits of the year number.

We count our years from the establishment of the first
human settlement on Mars, in 1997, so that Earth year
1997 is our year 1. This is necessary. We cannot use the
anno domini method of counting because our years are
different from Earth years and very soon the gap between
two similar numbering systems would have grown dread-
fully confusing.

My date of birth, for the record, was 3.68.06. That is to
say, I was born on the third day (which we call Tuesday, as
I said) of the sixty-eighth week of the sixth year of the
century. We omit the number of the century, but I was born
in the year 106. This is the martian year 132, and the date,
which everyone should know, is 4.22.32. If you wish to
translate into Earth years, then of course this is the year
2245 A.D.

We do our best, but there is no way we can avoid some
confusion when two groups of people are measuring the
passage of days and years but the days and years themselves are different. You get used to it, I assure you,
quickly enough, and by the time you reach Mars you will
be using martian reckoning and will all feel much younger!
As I mentioned, at twenty-six I am a grandfather. We
marry young on Mars. Partly this is because we need to
increase our numbers as quickly as we can, and partly it is
because our higher background radiation levels have a
genetic effect that accumulates through life. The younger
you are when your children are born, the smaller is the risk
that they will be damaged genetically. In fact, this risk is
much less than it may have been because we have adapted

quite well to the new radiation levels. Still, the age at which people can marry legally is eight. On Earth that would be fifteen. If it helps, you can remember that ten martian years are equal to 18.8 terran years.

Problems arise only when people need to use both systems rather than just one. Most of us live our lives entirely on one world or the other and have no interplanetary contact. People who need to maintain contact between the worlds can buy calculators with a time-date conversion function built into them, or buy dual-planet watches. Whichever they use, they must remember to set them in the appropriate mode, because such numbers as 4.3.45 on a document represent a valid date under either system. Legend has it that in the early days this fact caused a few sleepless nights among the fraud squads on both planets!

Now it is time for me to regale you with the travellers' tales I promised.

When you board a spaceship you never get a good look at it from the outside. You are shut in the ferry, strapped down during docking, and then transfer through a hatch from one vehicle to the other. The ferry approaches the ship nose-first, and the whole of it is hidden. You may have seen photographs or even holograms of a ship, perhaps even of this ship, the *Arcturus*, but in case you have not, today I will explain what it really looks like, and how it works.

It is shaped rather like a huge wheel. The wheel of a car, with a big, thick tyre on it, provides a familiar analogy. All the human accommodation is located inside the tyre part, around the rim, and once we are on our way the outer edge of the tyre becomes the 'basement' floor, with the other floors apparently above it, but in fact inside it from the viewpoint of the wheel. Some of the surfaces we use as floors while we remain weightless are then walls, and when you stand upright on the floor your body is aligned more or less along a radius of the wheel with your head pointing towards the hub.

There is a hub at the centre of the wheel, and it is connected to the 'tyre' by two spoke-like struts. These struts are used to carry cables and pipes that bring electrical power from the engine to the living accommodation, that connect the instruments in the control rooms to the sensors outside the ship, and that provide the small amount of power needed for the vast array of sensors that monitor the condition of the ship. These measure the strains on the structure, the temperature of the outer skin, temperature, humidity, and air composition inside the ship, as well as the operation of all shipboard systems. The struts also carry the ducting for the air conditioning, carrying air through the filters and purifiers that recycle it constantly, adding oxygen as necessary and removing surplus carbon dioxide and water.

Another ducting system removes sewage from the living quarters. It is carried to a treatment plant, where water is extracted from it, purified, and returned to circulation, leaving the solids as dry 'sludge' that will be ferried down to Mars when we arrive. We have an economy that is very basic! Martian farmers need as much organic soil conditioner as they can obtain, and human body wastes are valued highly. Those produced on the ship cannot be wasted, and since they are needed more urgently on Mars than on Earth, those accumulated during the journey to Earth are retained on board until the ship reaches Mars again.

The sewage has to be pumped. Once we are on our way, our simulated gravity acts outward, so that material moving from the circumference of the wheel toward the centre

must go 'uphill'. The lavatories themselves work as they
do on Earth or Mars, their contents being removed by
gravity, although they are very economical in their use of
water. This is not because the ship is or could be short of
water, but because it reduces the volume of material to be
pumped and the energy needed to dry it.

There is a much larger strut, or arm, mounted on the
hub and at right angles to the plane of the wheel. It
extends about one wheel-diameter from the hub. About
one-third of the way out there are the sensors used to
monitor the radiation to which the ship is exposed and the
loss of heat through the skin, and, most important, to
observe the positions of the stars by which the ship is
navigated. A further one-third of the way along, the arm
divides, with two smaller side branches leading to clusters
of pods, most of which contain water, but two of which
house the sewage and air conditioning plants respectively,
and one of which carries oxygen.

At its outermost end the main arm divides again, its two
branches leading to the much larger housings that contain
the engines. Passengers cannot go from the accommodation we use into the 'spokes' of the wheel, but access is
possible for members of the crew who may need to carry
out emergency repairs in the 'spokes' or in the main arm
during our journey. While the ship is parked in a planetary
orbit between journeys the whole of it is inspected, inside
and out, and necessary servicing is completed. The engines are replaced at regular intervals. Anyway, that is the
theory.

As you will have observed, the ship is fairly large.
There are four decks in the 'tyre', which measures about
twelve metres from the outer skin to the junction of the
'spokes'—the 'rim' you might say. The overall diameter of
the wheel measures sixty metres. You can work out from

that the circumference, which is 188.5 metres. Within the 'tyre' there are more than four hundred rooms, but only about half of these are used by passengers. The remainder include crew quarters, control rooms, stores, laundries, kitchens, and the hospital.

The 'tyre' appears firm, but that is because it is inflated. In fact it is semi-rigid and made from a material not unlike the carbon fibre invented in the twentieth century which gives it the great tensile strength it needs to withstand the stresses imposed on it.

We need large amounts of storage space because everything we use during the journey we must bring with us. In the early days many people speculated about the possibility of making spaceships self-sufficient by using conservatories to grow their own food, but also to purify air and water. The idea sounded plausible, indeed delightful, for every ship would have its own large garden. It was the size of this garden that showed conservatories to be utterly impractical. If they were to achieve all that was desired of them they would need to make each ship into a complete, self-sustaining ecosystem. This meant there would have to be more than just plants. The plants would need soil, with its full charge of flora and fauna. There would have to be pollinating insects, and therefore predators, plant consumers and larger carnivores, and with each addition the human payload had to be reduced. Something like an acre (half a hectare) would be needed to support each human. That was not the worst. If food was to be grown so intensively, two acres was about as much as one human worker could manage adequately if he or she had nothing else to do. That meant that each 'space gardener' could support one other human. Deduct the crew needed to operate the ship, and it became clear that the passengers would be heavily outnumbered! The price of a ticket would need to cover the cost of transporting the garden and gardeners. Finally, someone asked what would happen should a crop fail, say through diseases? The whole concept was abandoned. It was much easier to live as twentieth-century submarine crews lived for months on end, and carry food rather than the means to produce food.

The old idea may be revived one day, if ever we leave the solar system entirely and begin to explore the galaxy more widely, for then it will be impossible to carry sufficient stores. However, such starships would not carry fare-paying passengers, and by then we may have been able to automate much of the horticultural work.

The secret of successful space travel lies in security. Each essential system has up to three back-up systems, any one of which would be perfectly adequate in itself. They all work, and are tested regularly.

Power for the living accommodation can be provided by fuel cells. These electrolyze water, producing hydrogen and oxygen, which are then stored, the hydrogen in the form of a metallic hydride which avoids any need to cool the gas or subject it to intense pressure. When the cells need to generate power, hydrogen and oxygen are fed into them and recombined at electrodes, so generating an electric current with better than ninety-five per cent efficiency.

The fuel cells may be used, but they are needed only if the main source of power fails. That consists of a small thermal reactor housed at the far end of the main arm, between the main engine housings. It uses plutonium as fuel to generate heat, which is used to drive a turbine generator. One charge of fuel lasts for several years, and the reactor is very reliable. Being in space, of course, we cannot afford to carry the heavy shielding we would use on the surface of a planet, and so the reactor must be housed well away from the accommodation areas.

The main engines use a version of the 'ion-drive' principle. Each one consists of a small fusion reactor. As I am sure you know, fusion reactors have supplied most of the power used on Earth since early in the twenty-first century, but they are large. It was not until 2030 that a way was found to manufacture smaller versions.

An atom is composed of a large, electrically positive nucleus, surrounded by orbiting, negatively charged electrons, arranged in a series of concentric 'shells'. Usually, these charges balance one another so the atom as a whole is electrically neutral. If thermal energy (heat) is applied to the atom it may detach an electron from the outermost shell. In that case ionized atoms and electrons will exist independently of one another. The atoms from which electrons have been lost will carry a positive electrical charge, and in this state they are called 'cations', and the electrons ('anions') a negative charge. The generic name for a charged subatomic particle, of either charge, is 'ion'. Possessing a charge, the motion of an ion will be deflected by a magnetic field.

If a substance is to be fully ionized, so that atomic nuclei lose all of their electrons, very high energies are needed. You could expose it to intense gamma radiation or X-rays, but you would have to find some way of producing them in a controlled, sustainable fashion. Intense heat will do, but a thermal nuclear reactor is quite inadequate, and far too cool. It produces working temperatures of a few thousand degrees centigrade, but these are far too low. A fusion reactor, on the other hand, operates at temperatures of millions of degrees centigrade, which are adequate. It is why ion-drives had to await the fusion reactor, although designs for them had existed for many years.

A thermal, or fission, reactor, obtains heat through the natural decay of the nuclei of large atoms of certain elements, such as plutonium. A fusion reactor uses small atoms, and causes their nuclei to fuse together.

There is no physical substance that can contain gases at the temperature of a fusion reactor but, being fully ionized, they can be contained in a magnetic field, a 'magnetic bottle'. To convert the conventional power generator into an engine, all that was needed was a design of 'bottle' that included a hole—if you like, a 'neck'. The ionized waste products of the fusion reaction leave the 'bottle' through the hole, and much of the energy produced by the reactor is used to sustain its own reaction.

Power is transferred from the departing ions to the ship according to Newton's third law, which says that if a force is applied in one direction it must be balanced by an equal force applied in the opposite direction. Were the 'bottle' closed, pressure would be applied on all its sides. Since part of one side is removed, the pressure that would have been applied there is applied to the opposite side, and pushes the 'bottle', and thus the ship. The mass of an ion is extremely small, but the newtonian force is calculated as half the mass multiplied by the square of the velocity with which the mass moves, and the ejected ions move at a significant fraction of the speed of light. In any case, there are large numbers of ions involved.

The ion-drive engine is unsuitable for lifting a ship from the surface of a planet into orbit. It cannot deliver that intense, short burst of thrust needed during a launch. Its output is steady, but can be sustained for long periods. It allows a ship to accelerate slowly, but to go on accelerating until the required speed is attained. Our engines could have been used to accelerate us to a speed high enough to carry us right out of the solar system, had that been desired, but the acceleration was very gentle. It lasted for only a few seconds, and there was barely time for you to

feel it. Had you done so, it would have seemed like a
return of a little weight to your bodies. For that short time
you weighed only about one-tenth of what you weighed on
Earth. It was nothing like the huge pressure you felt during
launch.

The ship uses deuterium as its fuel. Deuterium, or
'heavy hydrogen', is plentiful, convenient, and safe, although 'heavy water' is moderately poisonous to drink. On
Earth it is obtained from sea water. It is convenient
because it weighs little. It is safe because it leaves no
harmful residue in tanks that may need to be repaired, and
cannot poison workers whose job it is to fill the tanks.
Even if it catches fire on Earth the burning gas, being
lighter than air, moves upward very rapidly, away from
people or structures it could harm, and the product of the
combustion (rapid oxidation) of hydrogen is nothing more
exotic than water. In years to come, engines may be
fuelled by helium-3, which is even safer, because it will
not burn at all, and in the reactor it produces almost no
radiation.

The main engines use only deuterium. They are started
when the ship accelerates, then shut down, to be restarted
to provide a reverse thrust to slow the ship in preparation
for parking. They work for only a few days out of a trip
that lasts months, and most of that time is taken in starting
and stopping them.

The thermal reactor uses plutonium and cooling water.
When, eventually, its fuel is depleted, the entire reactor
assembly is removed and replaced, with a fully serviced
reactor with a fully charged core. The old engine is
removed to a base on the Moon, where the old fuel is
extracted, reprocessed to recover the unused fuel it contains, together with such radioactive substances as can be

used in other industries and in medicine, and where the waste is stored.

Water, heavy water, and hydrogen are supplied from Earth and the ship carries a stock sufficient for the round trip. Water and heavy water are simply pumped into tanks and ferried into orbit, where full tanks are exchanged for empty ones. The tanks are not filled completely, of course, because the water of both kinds will freeze in space, and expand as it does so. The water pods on the arm contain ice while the engine is idle. Once the reactor is working, initially using liquid water inside its own jacket, the cooling pipes pass along the pods and so melt the water in them.

Hydrogen is transported slightly differently. It is pumped into tanks containing metallic fragments, looking rather like gravel, with which the hydrogen forms compounds that are stable until the tank is warmed gently, when the hydrogen is released. The weight of the metal fragments is much less than the weight of the strengthened tanks that would be needed were the hydrogen to be stored as a liquid, under pressure.

Either of the main engines would be sufficient to provide the propulsion we need. The ship has two only for safety. Were we to lose our drive completely during acceleration or deceleration we would be helpless, in a situation from which we could not recover, and we would have to be rescued. Although we have two engines, we use only one of them at a time. The second is started, so it can develop its full power quickly should this be necessary, but that is all.

The engine is used only to accelerate the ship to its cruising speed and to set its course. Once the engines have shut down, further minor course corrections are made by means of small, steering rockets on the main arm. These are fuelled chemically, by mixing two substances that ignite spontaneously when they are mixed. The substances themselves are stored in small tanks, and supplied in amounts that can be controlled very precisely. There are four similar small motors on the outer edge of the 'tyre'. These are used to make the whole of the living area rotate. The wheel turns on its hub. Then, when we prepare for parking, the rotation is stopped.

It is the slow rotation of the wheel that provides us with weight when acceleration ceases. It does complicate the geometry of the interior of the ship, but any alternative would be much worse.

At first, when we boarded, the wheel was stationary, in relation to the hub. This simplified the docking procedure, because we boarded directly into the 'tyre'. At first, therefore, we were weightless. Then the main engine was started and we began to move. The acceleration gave us some weight, briefly, and the direction we might have called 'down' was actually to the rear of the ship, the direction opposite to our direction of movement.

The wheel was set in motion as the engine was shut down, and it accelerated very rapidly to its final rotational speed. This is very slow. A rotation of a little more than one-tenth of a revolution per minute provides us with a weight equal to the weight we will experience on Mars. At last I am my own, proper weight again! It takes the wheel a little more than nine minutes to make a full turn, and the outer skin moves at 1.2 kilometres (about three-quarters of a mile) per hour. It is because the final speed is so low that it can be achieved so quickly. Slow it may be, but the rotation imposes huge stresses on the structure of the ship, which are even greater on the Earth-bound journey when the wheel must turn faster to simulate terran gravity. The concept of a rotating wheel is simple enough, but its design presented considerable engineering problems and might never have been possible without modern, semi-rigid materials which have great tensile strength.

As you know, we have been using martian time and martian dates since we boarded. That was the first step in acclimatizing us to martian conditions. The rotational speed of the wheel, providing us with our martian weights, is another step in that process. On the return journey, when the clocks and calendars use the terran system of reckoning, as I have said, the wheel will turn faster, and everyone will have their terran weight.

Speed, Economy, Comfort

From this point on our journey will be more comfortable. It was not always so.

The earliest migrant ships did not rotate. Passengers were weightless for the whole trip, and the trip itself was very much longer. It took nine months to reach Mars, and there were many casualties.

Those early ships were built for economy, not comfort, and certainly not for speed. They were small and they did not sail until they had as many passengers on board as could be crammed into their rooms. Weightless people slept zipped up in hammocks anchored at head and toe, so close together you could barely move among them.

The ships were powered by sails, huge sails kilometres square, so that although the ships were tiny, while they were parked in orbit they could be seen from Earth on a clear night. The sails were made from a very thin, reflective material, so that effectively they were mirrors.

They could be moved by means of tubes that inflated with gas, or deflated, according to signals from a central computer fed in turn from sensors, to ensure that they caught the light from the Sun at the required angle. The pressure of photons—and cosmic particles—striking them was enough to move the ships at what seemed great speeds in those days. They were much slower than ours, but with the sails set, they could accelerate at about five metres per second per second, so that it took rather less than ten minutes to accelerate the ship to the velocity needed to transfer it from Earth orbit to the trajectory that would carry it to Mars. Once the computer was 'told' the direction and speed required, it did everything else. In effect, the ship was entirely self-regulating, a complete cybernetic system that observed its own progress constantly and made all the necessary adjustments.

Like old-fashioned terran sailing ships, these spaceships could travel towards the Sun, or 'into the wind' as well as away from it, but for a quite different reason. In space, everything is in motion. You may think you stand still, as you think your planet stands still when you are on its surface, but all is motion. Earth orbits the Sun, travelling at 29.8 kilometres a second (more than 172,000 miles per hour). When you feel weightless, really you are falling freely through space, because 'weight' is merely the gravitational force that acts upon your 'mass'. Weight may alter, as gravitational force alters, but mass is constant.

If you are close to a planet, you will tend to fall towards that planet. An orbit can be explained by the newtonian laws of motion, which state that a body will tend to remain in a state of rest or uniform motion—motion in a straight line—unless some force acts on it to alter that state. If you are in orbit around the planet, two forces are operating on you. The gravitational force from the planet makes you tend to fall inwards, but this is balanced by your tendency to remain moving in a straight line, outwards and away from the planet, and that is how bodies can remain in orbit for long periods. They are not weightless, or more correctly massless, but in a free fall.

If you move away from a planet, you will tend to fall towards the Sun, so that if you are to control a spacecraft moving towards the centre of the solar system, its engines must be used to slow it, not to accelerate it. The sailing ships used solar radiation to drive them away from the Sun, angling their sails to achieve the amount and direction of thrust they required. They used the same radiation to slow them as they moved towards the Sun, and they used it in the same way. It was very simple, and usually it worked quite well.

Things could go wrong. The control tubes that moved the sails could fail, or the sails themselves could jam. Unless they could be repaired by humans working out in space, the ship was lost. One fully loaded ship suffered this fate, back in 2001. Its sails jammed while it was accelerating out of Earth orbit. It continued to accelerate, presumably until it was so far from the Sun that the amount of radiation it received became too small to accelerate it further. It missed its correct transfer path, and ended on a wide orbit that missed Mars completely. It left the solar system altogether and so far as anyone can tell, it should be well on its way to Alpha Centauri by now. Unfortunately, the first human star travellers died long ago.

Those old ships travelled by the cheapest route, and the cheapest route uses a Hohmann transfer orbit. The ship begins in orbit about Earth and accelerates so that its speed is 2.98 kilometres per second faster than that of Earth in its solar orbit. This additional velocity carries the ship into a higher orbit, the transfer orbit, which intersects the orbit of

Mars. When the martian orbit is reached, a further acceleration places the ship in the same orbit as Mars. The journey is so timed that the ship intersects the martian orbit just as Mars itself reaches the same point in its orbit, so the two meet. As soon as the ship is in the correct position in relation to Mars it slows down just enough to fall into an orbit around Mars. During the main part of the journey, the ship needs no propulsion, any more than we do, because it is in an orbit about the Sun. In fact it is falling around the Sun even though it is moving away from the Sun.

The advantage of using this route, apart from its low cost, is its simplicity. The navigational calculations require little computing power, and so the on-board computers can be small, light, and inexpensive. The disadvantage is the long journey time. The speed of the ship is predetermined by the orbit, and the journey from Earth to Mars took thirty-seven terran weeks.

During this time the humans on board were exposed to several dangers. Disease epidemics were one hazard that proved much more apparent than real. Susceptibility to disease varies from one individual to another, and crowding does little to increase exposure to infection. What matters is the maintenance of sanitary conditions, and efficient screening for disease carriers before embarkation.

People were exposed to variable amounts of radiation, but always more than the levels to which their evolution on Earth had adapted them, and when the winds of a solar storm enveloped a ship the radiation dose rose to very high levels. This increased susceptibility to quite ordinary illnesses by reducing the number of white corpuscles in the blood, but to some extent the effects were minimized because, to reduce the risk of fire, the air on board contained less oxygen than air on Earth—as does the air we are breathing. A low-oxygen atmosphere reduces the adverse effects of radiation exposure. All the same, you will be glad to know that our ship has a fully equipped hospital, with plenty of medicines for dealing with common ailments.

On the first few ships a small number of people died from radiation sickness while still in space. More usually, the effects were subtle. Any exposure is harmful, but the likelihood of serious damage seems not to be related to the dose in a simple, one for one, way. If the dose doubles the risk does not double as well. It is more likely that to double the risk you need to square the dose. This means that at very low radiation levels the chance of injury is very small. All of us evolved in a radioactive environment, after all, because Earth itself is radioactive, and we have survived.

When exposure increases not everyone is affected equally. Just as some individuals are more susceptible than others to certain physical disorders, so some are especially vulnerable to radiation damage. The radiation spared the more robust, but among the injured the consequences were not dramatic. They did not suddenly fall victim to some dread illness. It was just that they tended to die younger than the others. This selection, for that is what it was, continued after people arrived on Mars, for the martian background level is higher than that on Earth. After a few generations most Martians were more tolerant of radiation than Terrans. This had interesting evolutionary consequences.

So far as individuals are concerned, the risks of low-level radiation are often exaggerated. During the Second World War, the instrument panels of aircraft contained instruments whose dials were painted with radium-based luminous paint. At night they illuminated the whole cabin, but they emitted radiation by day as well. Scientists pointed out that the crews were being exposed to unacceptably high radiation levels, but it was felt that the benefit to airmen of knowing their position, airspeed and altitude was greater than the harm being done to them by their own instruments. In any case, the hazards of combat flying at that time were such that statistically it was unlikely that crews would live long enough to suffer illnesses induced by radiation. All risks are relative! A rather similar argument might apply to space migrants.

There was another, much more immediate problem. When you are weightless your bones lose calcium. This was discovered very early in the history of manned space flight. When you are within a gravitational field once more the loss ceases and over a short period of time the calcium is restored. It was hoped that during prolonged periods of weightlessness the calcium loss would cease, that a point would be reached beyond which the skeleton retained such calcium as remained. It was not so. Calcium loss continued indefinitely, with the promise of eventual severe damage to the skeleton. Today we simulate gravity by rotating the ship, and that is one of the reasons why we do it. The early pioneers remained weightless for the entire journey, for about nine terran months, and there were some who arrived permanently crippled.

The journey was hard and uncomfortable. Your body lost calcium. You were exposed to high, and sometimes dangerously high, levels of radiation. There might be disease. There might be a failure of some ship system that would lead to catastrophe, perhaps in the form of a slow death in a ship drifting out of control and beyond the reach of possible helpers.

Can you imagine what it must have been like? Remember that everything was done as cheaply as possible. There were no luxuries.

You would arrive at a ramshackle spaceport, to be crowded into a dirty, badly furnished lounge, where you were kept waiting, sometimes for days. Then you boarded the ferry for the launch. That was the best part of the journey, for the ferry was subject to civil aviation safety standards. It was reliable, safe, and came close to being comfortable.

Once aboard the spaceship you were herded to your quarters, to the part of the ship in which you would spend almost all of your time during the voyage. There was barely room to move among the people, so crowded was it. You slept zipped up in a hammock, secured at head and toe, and when you were out of your hammock you were inclined to drift, so you stayed 'in bed' much of the time. If you wanted entertainment you supplied your own, but exercise was impossible. Even the old, wooden, migrant ships of Earth had decks where now and then even the poorest passengers could walk in the fresh air. Spaceships had no such facilities and there is more to exercising when you are weightless than just walking around. You need plenty of space and you should have proper equipment.

Each ship carried medical software so that its computers could diagnose ailments and prescribe treatments, and ships were provided with the barest minimum of medical stores, but the shipboard medical facilities could not cope with an outbreak of illness affecting more than a few people at a time.

Food was poor, and there was never enough of it. For a time most passengers suffered from space sickness. You know something about that, but in those days it went on for much longer. It is a form of motion sickness, caused by the disorientation of your senses when you are weightless. It passes, but the imposition of a gravitational field makes it pass much more quickly. Those old ships did not rotate, and they accelerated for only a few minutes, while they moved out from their Earth orbit. Until you arrived, if you

did, you weighed nothing. When people recover from
space sickness they are hungry, and on the old ships they
stayed hungry.

If physical conditions were bad, psychologically the
ordeal was no easier. Those pioneers possessed nothing
more than the clothes they wore and whatever they could
carry as hand luggage. They had sold everything just to
pay for the ticket and to buy, in advance, a place on Mars
in which to live and work. Temporarily they were desti-
tute. They knew very little about the world to which they
were going, they were totally unprepared for the squalor in
which they had to live. Confused, often frightened, hungry, and more or less sick, they fell through the sky packed
tightly together with little more than their dreams to
sustain them.

Why did they do it? The question has two parts and
neither is difficult to answer. They decided to travel in the
first place because of their dreams, for the challenge
perhaps, for the opportunities their new world afforded. It
is easy to understand how families must have felt as they
sat in their homes on Earth adding up the value of those
homes, and all their possessions, to discover they could
afford to make 'the big jump'. That is what they called it,
'the big jump'. In a fever of excitement they would apply,
be interviewed, and after that matters were taken out of
their hands. They did not sell their own goods and use the
money to buy their migration documents. The agencies
took complete control, through papers with which all
migrants signed over their goods. They were told it was
more convenient to operate in this way, and there was no
choice. The agency did the selling, when prices were
right. It was simple for the migrants, and profitable for the
agencies.

They were told of the wonders of Mars. They saw

pictures of spectacular scenery, of laughing people, and of space sailing ships whose beauty was undeniable. The huge mirrored sails looked clean and who would imagine that the interior of the ship was less wholesome? They were told of the skill of the crews, the reliability of the computers, and of the months of quiet relaxation they would enjoy before beginning their new lives. It was not until they had boarded that they began to suspect that all was not as it had been painted. By then it was too late.

The only people who knew the truth about the ships were either dead or on Mars, and there was very little direct communication between Martians and Terrans in those days. Stories did leak back to Earth from time to time, but they spread as rumours, and were denied.

You may wonder why the terran authorities did not act to protect migrants, but it is doubtful whether even they knew the whole truth. Even now it is written into martian, not terran history. Agreements between individuals and the spacelines were legally binding and negotiated freely. There was nothing improper about them. The ships themselves were inspected, and always satisfied the inspectors. Inspections took place while the ships were parked in orbit, with skeleton maintenance crews aboard. The living quarters were spotless, the facilities appeared adequate, the ships were spaceworthy.

Perhaps some officials knew, or suspected, that the ships were overloaded, but overloading is not the problem in space that it is on Earth. Spaceships will not sink, or overrun their runways, or wallow out of control in heavy seas. Payloads are critical when launching from a planet into orbit, but once in space they are much less so. A heavier load means no more than a slightly longer period for acceleration. If the ship is sail-powered it means only that the sails must be set a few minutes sooner.

There may have been a political factor, too. Interplanetary migrants would never again vote in a terran election. The moment they signed over their property they resigned their franchise, and politically they ceased to exist. Civil rights have a lot to do with votes.

Today, you will be relieved to learn, the authorities on both planets collaborate to impose very rigorous standards on interplanetary vessels, covering the comfort of passengers and crew as well as their health and safety. Infringements can be reported by passengers on either planet, and conditions on board are monitored constantly by computers and sensors installed officially, with which no one can interfere without detection. Should there be a serious infringement of the safety codes, not only is the spaceline penalized heavily, the ship is confiscated.

Travel is safer today. Sailing ships looked beautiful, but they were as dangerous in space as their romantic predecessors were at sea. The *Arcturus* may be old, but its engines are reliable and we move under our own power. That allows us to reach higher speeds, a shorter trajectory, and so a much shorter journey time than we would need under sail. The old Hohmann orbit was abandoned long ago, except for unmanned cargo ships.

It may interest you to know that the sailing ships travelled at 117,720 kilometres per hour (110,435 miles per hour) relative to the Sun. That represents the speed of Earth in its solar orbit, plus the additional speed needed to enter the transfer orbit. At this moment, we are travelling at about 200,000 kilometres (124,000 miles) per hour.

Navigation in space presents a few problems of its own, because the reference points are not those that might be used on a planet. It is not quite so simple as going from a 'here' to a 'there', because everything is moving in relation to everything else. The computers do all of the work.

The job of the crew is only to tend to the needs of the passengers and, most of the time, look decorative.

The orbital path the ship must follow is plotted in relation to two factors, the stars and radio beams. The relative positions of stars so remote that they appear motionless are calculated, by computer from standard charts, at five-minute intervals throughout the voyage. Sensors on the main arm of the ship observe the stars, their observations are fed as signals to one of the computers, which compare the observations and the maps.

Radio signals are transmitted from both Earth and Mars, and are received by the ship, using a directional aerial. The ship cannot simply 'sail the beam'. The planet bearing the transmitter is moving. If the ship sailed straight towards the transmitter it would describe a curved path that carried it into the orbit of the planet, but behind the planet. We would have to chase Mars around its own orbit. That is possible, but wasteful. What we aim to do is to intercept the planet, so that it and we reach a predetermined position in its orbit together. Thus the direction of the signal must change at a particular, calculable rate. This rate is fed into the navigation program and another computer compares it with the actual direction of the signal that is received.

The information gathered by the sensors and matched against the program ends at a 'committee' of three microprocessors linked to the systems that control the ship. If all three send the same signal to the controls, all is well. If there is disagreement the microprocessors check their data and their calculations and if it transpires that one of them is malfunctioning it is isolated until it can be repaired. This may sound complicated, but technically it is not difficult, and the navigation of a spaceship demands very little in the way of computer memory. Almost any domestic computer has sufficient computing power. Thus the microprocessors determine when engines shall be fired and shut down, when rotation shall begin and end, when course corrections shall be made. It is they that take the ship out of orbits and place it in orbits. It is very unlikely that the entire system could fail, but mainly to reassure the humans, they display what they are doing on instruments in the control rooms. The crew can observe the condition of the ship, its life-support systems, and its progress toward its destination. If the captain thinks it necessary he can activate a manual override and sail the ship himself.

I have said nothing of the conditions the pioneers found awaiting them when they landed on Mars. That must wait, and before I can describe Mars as it was then I must tell of the way the transformation of the planet altered it from the primeval state.

I leave you with a curiosity of spaceships. If you stand on the surface of a planet, you can imagine a line from your centre of gravity to the centre of the planet. If you stand so that the line passes more or less down the centre of your body, you are standing upright. However, the surface of the planet is curved, so that another person standing upright beside you actually is inclined away from you. The difference is so slight you cannot perceive it, and it can be cancelled out entirely by irregularities in the ground surface itself. On this ship, however, where the ground surface is regular and which is tiny in comparison with a planet, you can see the curvature clearly. We are on the inside of the wheel with gravity acting in an outward direction, instead of on the outside with gravity acting inward, so the curvature is concave instead of convex, but that is the only difference. All the floors are curved, and if you look along a corridor you can see it rise in both directions, in front of you and behind you. Yet as you walk along the corridor, there is nowhere that you move down-hill or uphill. It is perfectly level where you are or, to put it another way, you are always at the bottom of the slope. A picture on the wall looks straight while you are in front of it, but if you are some distance to one side it looks crooked. If you should have a spirit level about you, that will confirm that everywhere the curved floor is in fact level! You will never see this again so clearly

You can have fun with it, what is more. If two of you stand facing one another, just a little way apart, close your eyes and stand as upright as you can, then open your eyes, you will find that you seem to be leaning towards one another, like a pair of music-hall drunks. You must look quickly, because at once you will try to correct yourselves— by leaning outward. Don't fall down!

Greenhouse
and Daisies

We must be nearly half way to Mars by now. You can almost work out how far the ship has travelled by the conditions on board and the distress the passengers are experiencing. By the half-way mark, the acclimatization process is far advanced and conditions on board resemble quite closely those the travellers will encounter inside buildings on Mars.

Our meetings have become more occasional, more desultory. They were never formal, or arranged to a timetable like a course of lectures. They just happened, more or less spontaneously, whenever a few of us found ourselves together with time to spare, usually after the evening meal. We just sit and talk, although I like to make a few notes in advance, or at least to think through the ground that remains to be covered, so that our discussions have some structure. I enjoy them, as much for myself as for the benefit they may bring to my friends. More and more I find myself exploring my own thoughts, recalling as much as I can of our history, trying to understand why events fell out as they did. I have a strong feeling that our future will be different from our past, that we have reached the end of an era. It is a time for introspection, then, for accounting.

I think our talks help to relieve the boredom. There is not much to do on a space voyage. You can gaze into space, as space is depicted on the screens, much as passengers once gazed vaguely at the sea from old-fashioned terran ships, but such entertainment palls rapidly. Official entertainments are provided. There is music and drama, recorded of course, there are old films to watch and panel games in which passengers participate. Anyone with a real, or more usually imaginary, talent is pressed into service. A kind of frenetic jollity is simulated by the crew. They talk too quickly, too loudly, and with their voices pitched just a shade higher than is customary, but they are not very good at it. They, too, grow bored, and it is all too much of a strain. After a week or two the novelty wears off, everyone relaxes, and we get to know one another.

Then there are the computer entertainments. These consist mainly of adventures into which players can enter, either to leave again after a short exploration or to remain, becoming more and more engrossed in the fantasy presented to them. They are not to everyone's taste.

Games invented centuries ago come into their own. People play board games of ancient and honourable lineage. Chess and Go are very popular. You can even find people doing jigsaw puzzles. Card games attract large numbers of passengers, and bridge is the game that divides us into three groups. There are those who play, those who are being taught to play, and those like myself who can never quite grasp the purpose of the game and find even star-gazing preferable.

In general, though, the meetings have become fewer not because people are losing interest, but because they are sleeping a good deal. Even I feel lethargic, though I am affected less severely than the Terrans. Over the last week or so, little by little the air pressure in most parts of the ship has been reduced. Altitude sickness has become common and it can be quite distressing. People have blinding headaches, lose their appetites, and even when they feel well the slightest physical effort exhausts them.

The oxygen content was lower than that of ordinary terran air from the time we boarded. That is to say that the air we breathe contains rather less oxygen than a similar amount of air would do on Earth. Mainly for reasons of fire prevention, spaceships always maintain such a low-oxygen atmosphere. This has nothing to do with the pressure change. Provided the proportion of oxygen in the air remains high, the air pressure can be reduced to very low values and it is still possible to light fires. Terran mountaineers use fuel-burning stoves to heat their food, even close to the peak of Everest. If the pressure remains constant, however, a reduction of the oxygen content can be achieved so that although nothing will burn, people can breathe perfectly well.

So far the effect of the pressure reduction is similar to that of moving from sea level to an altitude of about 5000 metres (16,400 feet) on Earth. That is rather higher than Mont Blanc (4810 m, or 15,800 feet) in the European Alps, or Longs Peak (4345 m, or 14,255 feet), a popular climb in the Rocky Mountain National Park in Colorado. Tibetans would feel perfectly at home in air as thin as this, and so will we in a little while, but it takes time to grow accustomed to it if you have been living at or near sea level.

For safety reasons, as the pressure has been reduced the proportion of oxygen in the air has been kept constant, so that it contains less oxygen than does the air we are used to breathing and even less than it did when we boarded. Because it contains less oxygen, our bodies are limited in the raw material with which we oxidize carbohydrates to obtain energy. That is why we sleep, and why quite modest exercise tires us. The human body compensates by using more efficiently such oxygen as is available to it. It manufactures larger amounts of haemoglobin, so it can abstract a larger proportion of the oxygen we breathe. Once that adjustment is complete all is well again. It is a reversible kind of adaptation. Those of us who were born on Mars have blood with a high haemoglobin content from birth, but when I went to Earth it soon fell.

I said the change is like what a Terran would feel following a move to a higher altitude, but in fact it is easier if you travel between the planets than it would be were you to remain on either one. Gravity helps. When you go to Mars, the reduction in gravity means your body has less work to do just when it feels deprived of energy. When you go to Earth, for a little while you have more energy, but your weight increases, and so you must work harder.

While all this has been happening, the composition of the air has been changed. On Earth the air consists mainly of nitrogen and oxygen. On Mars it consists mainly of carbon dioxide and oxygen, although there is an increasing amount of nitrogen now that life is established and thriving in martian soils, cycling it from the soil to the air. You can tell when the composition changes by the smell. At first, martian air has a slightly pungent smell—the smell of the carbon dioxide—but you cease to notice it after an hour or so.

What you do notice, and it takes a little getting used to, is the way our voices change. Terran deep-sea divers develop high, squeaky voices as soon as they begin breathing a helium-oxygen mixture. We experience the opposite effect. The carbon dioxide-oxygen mixture we breathe gives us deep, growly voices. Tenors and sopranos are prized more highly on Mars even than they are on Earth, because they really are rare. Most Martians sing and speak in the lower registers.

We have been exposed to martian temperatures since the journey began. They are not much different from the temperate regions of Earth, but we tend to keep our buildings at rather lower temperatures and if we feel cold we put on warmer clothes. It saves energy. There are no fires—indeed, no naked flames or even sparks—in martian buildings, for reasons I will explain in a moment. The temperature on board, and in most martian buildings, is around 13°C (55°F). That is about the temperature we expect on a spring day where I live. In summer it can be much warmer than that, and our buildings feel pleasantly cool as you enter them. I have known the temperature to rise above 27°C (80°F) in summer. I live close to the martian equator, of course, but even so it would be quite wrong to think of Mars as the icy desert it once was.

As soon as everyone is used to the reduced air pressure, the pressure will begin to fall again, more slowly until it reaches a value equal to that you would experience at an altitude of about 7000 metres (23,000 feet) on Earth. This is high, the air is very thin indeed compared with at sea level, but people can and do adjust to it provided the transition is made slowly and at a time when they can rest as much as they wish. The medical examination detects most of those individuals with a weakness of the heart, circulatory or respiratory systems who might have difficulty, and we can deal with emergencies on board. We have pressure chambers as large as comfortable rooms. If necessary, people can be moved to them and the air pressure increased gradually, then reduced again more slowly than it was the first time. We have ample time for acclimatization, and we use all of it.

The final stage in acclimatization, in which we will all adjust to the atmosphere we will breathe inside martian buildings, has to take place on Mars itself. Then the pressure will be reduced still further, but the oxygen content of the air will be increased until it becomes very oxygen-rich. We can breathe such air without difficulty—but not on board ship. Remember that the ease with which materials burn is related not to air pressure but to the amount of oxygen that air contains.

Naturally enough, it is the reduction in air pressure that provides the main topic of our somewhat breathless conversation. Once the symptoms of altitude sickness disappear, the further changes occur gently and cause no distress to most of us, but everyone has been told what is happening. People seem surprised either that the air is so thin, or that it is not thinner, and all of them ask why it is we maintain such low pressures in our buildings. If buildings are sealed hermetically, which they are, why do we not provide ourselves with more comfortable pressures?

The first part of the answer, of course, is that these pressures are perfectly comfortable so far as we are concerned. We would not wish them higher. Even if we did, it would not be advisable. For one thing, it is convenient to be able to move out of buildings and back into them again without having to experience drastic pressure changes that could cause us considerable physical harm. It would do nothing to enhance the quality of our lives if we had to dress in cumbersome, pressurized space suits or go through pressure chambers just to walk in the garden.

Our buildings are hermetically sealed, not to contain a higher air pressure indoors, but to keep separate two different kinds of atmosphere.

From a terran point of view, the air pressure is very low, about one-sixth of the pressure at sea level on Earth. It is as though you were at an altitude about twice the height of Everest. We can breathe in such thin air only because, in our buildings, the air contains sufficient oxygen. Outside, the air is still mainly carbon dioxide. One day we will have an oxygen-rich atmosphere like Earth, but that day is still fairly distant. So when we go outdoors we need to take our air with us. This is less onerous than it sounds. You need no special clothing, except for a mask, fed from a small bottle you carry on your back. Some people prefer a helmet, which they find more comfortable than a mask covering the nose and mouth. It is a matter of personal preference. It makes conversation a bit difficult, but even wearing a mask or helmet, you can be heard if you speak clearly.

I should say something about the breathing apparatus itself. At first, in the very early days when our atmosphere had almost no oxygen, people used bottled air, fed through a tube to the mask. Then martian scientists invented a perfluorocarbon polymer elastomer, a plastic material made from perfluorocarbon. Perfluorocarbon had been known on Earth for many years, as a liquid. It has the property of forming a temporary, reversible bond with oxygen, much as haemoglobin does, but its oxygen-carrying capacity is far greater than that of haemoglobin. On Earth they experimented with it as a substitute for blood, to be used in transfusions. An animal that breathes air with lungs, such as a mammal, can breathe perfectly well when submerged in this liquid.

The martian innovation was to make a plastic solid from the liquid, and so to construct a breathing apparatus that requires no bottled air. It could work as soon as the martian atmosphere contained an amount of oxygen that was significant, but still far too little to allow a human to breathe unaided. When layers of perfluorocarbon membrane are laid one on top of another, then rolled like a Swiss roll, they will let oxygen pass but not other gases. If the roll is attached by one end to a breathing mask and by the other end to a small, battery-powered, electrically-driven air pump, the pump will force ordinary martian air through the roll, but the person wearing the mask will breathe almost pure oxygen. The advantage of this lies in the fact that modern electric batteries are small, lightweight, and last for a long time. They are much easier to carry than an oxygen bottle, and last longer.

Maintaining the same pressure inside and outside is more than a matter of convenience. All martian buildings have a porch at every entrance that acts as an air lock. They are very safe and reliable, but inevitably a hermetic seal carries with it the risk of catastrophic failure. Imagine what would happen if an air lock were to open both its doors at once, and then jam. It is all very well to say that cannot happen, but impossible things have a habit of occurring now and then. Suppose the people in the building were used to high—say terran—air pressure, and suppose the building were small, like a private dwelling, and so contained only a small volume of air. Air would flow out of the building. There would be an almost instantaneous drop in pressure. People would collapse and possibly die before anyone could reach them. As it is, the failure of an air lock will cause oxygen-rich air to flow from the building, or carbon dioxide-rich air to enter, depending on the tiny actual differences in pressure that are bound to exist across any air lock. There would be ample time to detect the fault and deal with it, even if this meant advising the occupants of the building to don breathing masks temporarily, and that would be necessary only if the fault were such that the air lock could not be closed at all. If the air lock remained open for a long time, oxygen-rich air would flow outward near the top and heavier carbon-dioxide would flow in close to ground level, and the building would fill with an unbreathable mixture from the floor upward. It really is very safe, at least from that point of view.

As a fire hazard, the air in our buildings is extremely dangerous. That is why we have no fires, no naked flames, nothing that could make a spark, and all electrical wiring is sealed within fireproof sheathing. Twentieth-century smokers would find Mars most uncomfortable. Indoors they could not be permitted to smoke, and outdoors their smoking materials would refuse to burn!

So people adjust to martian air, as they adjusted earlier to martian gravity and martian time, and as they have, without noticing it, adjusted to martian food. There had been complaints about the cuisine on board that receive the vaguest of replies. What I know, but the Terrans apparently do not, is that the cuisine is excellent, but martian. All of us are what Terrans would call vegetarians, but that is a matter I will explain later.

Having said so much about martian air and martian temperatures, I must explain how our planet was transformed to its present, delightful state.

The Foxe rockets began to arrive around the end of the

terran year 1985, you will remember. Most people had forgotten all about them. There was a flurry of excitement when the first few were launched in the spring of that year, but it soon subsided. The launching of rockets was a routine event by then, attracting little attention, and these launches might not have been reported at all were it not for their destination, their cargo, and their purpose. Even then, one rocket launch is much like another and it is difficult to report the tenth, far less the thousandth, with the spontaneous enthusiasm generated by the first. A rocket fires, rises into the air with a great white vapour trail, and vanishes from sight. There is not much that even the most inventive journalist can say to make it sound exciting.

By summer the story of the rockets to Mars was relegated to the bottom of inside pages and by Christmas it had no news value whatever. The people of Earth were concerned with their traditional pursuits. They celebrated the festival, they ate, they drank, they read about eating and drinking, they followed tales of scandal involving the famous, they quarrelled, they gambled, they forgot all about space. When the first rockets arrived the fact was not reported. Indeed, it was not observed, for not even the most powerful telescopes could have seen it and the rockets themselves carried no device that would have notified anyone of their arrival. They crashed in obscurity.

As each one crashed and was destroyed, it released its cargo of CFC as a gas into the martian air, and such was the pattern of atmospheric circulation on Mars that within a few days the gas was distributed evenly over the entire planet. A thermal blanket began to form immediately and as rockets continued to arrive, so the blanket thickened.

It was not until April, 1986, that the results were seen from Earth, and then they were reported—incorrectly.

Most of the martian northern hemisphere was covered by a thick pall of dust and all features disappeared. Clearly, the largest dust storm ever seen was in progress. Immediately, the popular press produced suitably blurred photographs to accompany stories about the great clouds of dust raised by the impacting rockets. No doubt each rocket did raise some dust as it crashed, but this had nothing whatever to do with the storm. The true story was much more interesting, but much more complex.

Do you remember my description of the martian dust storms? They usually begin in spring, in the southern hemisphere. As the ground is warmed by the Sun, the carbon dioxide hoar frost begins to sublime. When a substance changes from solid to liquid, or liquid to gas, energy is needed to effect the transition by breaking the bonds between molecules. That energy may be taken from the air or surrounding materials, so cooling them. It is why it feels colder when ice is melting. The latent heat of sublimation chills the ground immediately beneath the frost, so that the ground that was covered by frost is much colder than adjacent ground where there was no frost. This temperature difference caused turbulence in the air close to the surface, the turbulence caused dust to be picked up, and so the storm began. The dust in the air was also warmed by the Sun, warmed the air, and increased the turbulence, which meant still more dust was lifted from the ground, and the carbon dioxide itself, being a 'greenhouse' gas, contributed to the warming. The process contained what scientists would call a large positive feedback—it fed on itself and grew.

The CFCs increased the effect. They retained heat in the air before the ordinary spring warming began, and so started the carbon dioxide sublimation early. Then spring arrived, the usual annual warming was added to the effect of the CFCs, and sublimation proceeded much more rapidly than usual. The total area covered by solid carbon dioxide shrank very rapidly, so that the air turbulence was much greater than in previous years. The dust storms were correspondingly larger. They began earlier, were more intense, and, more important, they lasted for much longer. For the whole of that martian spring, summer, and autumn the storms raged over virtually all of the southern hemisphere. Winter was well advanced before the dust began to settle, and then it took some ninety days before the air was clear again. In all, the storms had lasted for sixteen terran months.

The rockets fell everywhere, quite indiscriminately, and the CFCs were distributed throughout the atmosphere—in both hemispheres. As the storms died down in the south, so they began again in the north. The period of clear skies was short, and soon the other half of the planet was hidden.

By the time the storms abated, the average ground-level winter temperature on Mars had been raised by a full 30°C, and on the low northern tropical plains atmospheric pressure had risen to about twenty millibars, equal to the pressure in the middle of the terran stratosphere. Of course, the effect of those first rockets was not the end of the change. They continued to arrive, to crash, to release their CFC cargo, and temperatures and pressures continued to rise.

The rise in temperature was sufficient to sublime all the solid carbon dioxide from the surface, and to keep it in the gaseous phase. Before long carbon dioxide hoar frost became a phenomenon of the past, and no Martian has ever seen it. Water near the ground surface melted, and evaporated instantly. The evaporation chilled the ground, and made a small contribution to air turbulence. It was the water vapour and carbon dioxide in the atmosphere that raised the pressure. At high altitude, some of the water vapour condensed around dust particles to form clouds, and there was real rain, in the form of weak carbonic acid, for the water dissolved carbon dioxide as it fell.

In previous years the storms abated by late summer, the dust settled, and the water vapour and then the carbon dioxide were frozen out of the air in winter, producing the icecaps. In the middle of winter, the northern icecap used to extend to about latitude 60°N, occupying a substantial area of the hemisphere. The southern icecap was even larger in winter, although it used to be very small in summer.

In that year it was different. Because the storms lasted well into the winter, the dust settled on top of the ice, rather than settling first, and therefore lying beneath the ice. This altered the albedo of the planet. The albedo is simply the 'brightness' of the planet as seen from space, the extent to which it reflects sunlight. White areas reflect a great deal of sunlight, and so have a high albedo, while dark areas reflect much less light, and have a low albedo. Sunlight that is not reflected is absorbed, and radiation that is absorbed warms the surface that absorbs it. In the winter following the first storms, the icecaps formed as usual, but they were beneath the dust, and dark in colour instead of being white.

Nor was the dust its usual martian colour. The destruction of the rockets had released into the air soot-like particles of hydrocarbons derived from impurities in the CFCs. No one has ever discovered how the impurities became mixed with the CFCs. Personally, I think they were added deliberately, and that they consisted of nothing more exotic than common terran soot. Whatever their source, these particles were distributed throughout the atmosphere, and they darkened the dust with which they mixed.

Their effect was minimal, but it served to conceal for a time the darkening that produced a much larger effect. From the earliest days of planetary exploration scientists had warned of the dangers of contamination that might be caused by biological organisms carried inadvertently from Earth. The fear was that such organisms might proliferate in their new habitat and distort any existing ecosystem to such an extent that it would be impossible later to reconstruct that system. No one could ever know what the planet had been like before the contamination.

When the team preparing the Foxe rockets were planning the changes they hoped to make in the martian environment, two things became clear. In addition to the chemical cargo it would be necessary to send some substance that would reduce the albedo of the planet, and the only suitable candidate would be a terran living organism. At the same time it was obvious that the presence of such organisms on the rockets would have to remain a secret. Objections to the climatic modification had been overcome, more or less, but the deliberate injection of terran species might have been too much for terran scientists to accept.

Expeditions to Antarctica had found the continent to support some plant life, even deep in the interior. There were areas of dry desert, free from ice, where a few species of algae were able to survive, and some of those strange algae were black in colour. Their colour and their ability to thrive in an environment as harsh as that of Antarctica made them prime candidates for migration to Mars. Quietly, unobtrusively, samples of antarctic soils containing them were collected and found their way to southern India. Carefully the algae were separated from the soil and studied. Their nutritional requirements were analysed, and refrigerated rooms were built in which they were cultured. Algae that grow in soils, such as these, can survive long periods of dessication. Some have been known to start growing after remaining dormant for half a century under conditions of extreme dryness. As a group, algae reproduce by various means, but most, including the antarctic algae, do so by means of spores. When the time came, spores were collected, placed in suitable containers, and sealed into the rockets.

The packages of spores were far from pure. They contained spores of other algae, and of such bacteria as had been associated with the algae in their original antarctic site, and all of these species were disseminated together. On Mars they settled, grew, reproduced, and within a fairly short time had covered large areas, obtaining all the mineral nutrients they needed from the martian soil and carbon dioxide from the martian air. When the storms cleared and the planet was seen to have darkened, hydrocarbon impurities were blamed. By the time critics had performed the calculations necessary to show that the rockets could not have carried the millions of tonnes of soot needed to produce the darkening that was observed, many more rockets had departed, the process was far advanced, and it was much too late to reverse it.

Quite apart from their immediate effect on the albedo of our planet, the life that had been introduced began the formation of fertile soil and the rearrangement of the atmosphere.

Where once they reflected radiation, and contributed to the general cooling, the icecaps began to absorb it, and contributed to the warming. They did not warm sufficiently to melt the ice, but the polar temperature did not fall so far as it used to do, and in spring the cap melted sooner. Of

course, the algae released by the rockets were not confined
to the polar regions. They fell everywhere, and grew
everywhere, so that the albedo of the entire planet was
reduced substantially.

The second season, and the continuing arrival of rockets, saw more storms. They were less intense than those of
the first year because no carbon dioxide had frozen and
there was no rapid sublimation to cause extreme air turbulence. Nevertheless, the melting and evaporation of water
had some effect, the algae continued to grow, and once
again the winter icecaps were darker in colour. They were
not covered entirely by dust this time, but sufficient dust
had remained in the air in early winter for it to have mixed
with the ice, making it 'dirty'. So the process continued,
year after year, and each year saw what humans would
regard as an improvement in the martian climate.

By the time the first humans landed on the planet,
fourteen terran years after the first rocket, the northern
icecap was very small indeed during the summer, and
during the southern summer the icecap at that pole
disappeared entirely. Latitude for latitude, Mars was cooler
than Earth, but in the tropics temperatures throughout the
year were not much different from those of the central part
of North America, or of Eurasia. The planet was definitely
habitable, and liquid water had begun to flow.

The storms were intense and widespread, but storms like
them must have occurred before. Planetismal impacts must
have had rather similar effects, and we know that those
effects did not endure. Eventually, after a few years perhaps, the climate returned to its former state. That these
storms were different, that the warming was sustained, was
due partly to the CFCs, whose 'greenhouse' influence
continued until the gases themselves began to disappear
from the atmosphere. That did not happen until a martian

century or so after the last of the Foxe rockets arrived. Mainly, though, it was due to the proliferation of the algae.

It was not necessary to maintain temperatures and pressures indefinitely, for the arrival of humans changed the situation yet again. The people who came began at once to sow more plants in the new martian soil, together with the soil micro-organisms that are needed to process soil minerals and make nutrients available. This was perfectly possible, with a few simple modifications of standard terran agricultural and horticultural techniques. They brought with them the sewage sludge accumulated during their voyage, dried and compressed organic soil conditioners, fertilizers and seed. Once the organic materials were mixed with martian soil, a martian version of agriculture could begin.

The algae and bacteria that had arrived with the Foxe rockets had adapted rapidly to martian conditions. Among single-celled organisms evolution can proceed very rapidly indeed, because reproduction is so rapid. They comprised the first truly martian species, for by the time humans arrived they were markedly different from their terran ancestors.

Temperatures were ideal during the spring, summer, and autumn, and those seasons were long enough to produce two harvests in the first year. The atmosphere contained little oxygen, but plants require little oxygen. What they need is carbon dioxide, and Mars has never been short of that! Indeed, on Earth it is often the shortage of carbon dioxide that limits the growth of plants. Martian soils were induced to supply mineral nutrients without much difficulty. On Mars, plants grow faster and larger than they do on Earth, mainly because of the abundance of the raw material for photosynthesis. Water was available, and crop pests were absent. These first settlers were not the miserable migrants who followed later. They came well prepared, and they knew that it was not only food and fibres their crops must provide.

These crops, grown outdoors, were not the principal source of food for the settlers, and the provision of food was not the main reason they were sown. Food crops were grown much more efficiently, and even faster, in glass-houses, using nutrient film techniques. The outdoor crops were grown primarily to build on the work of the algae, and so to accelerate the transformation of the planet.

On a planet that supports life, conditions are maintained to a very large extent by the living organisms themselves. Provided conditions are produced in which life can begin and become firmly established, and provided there is no major interference from outside, micro-organisms, plants and small invertebrate animals are capable of 'taking over' an entire planet and altering it to suit themselves. After all, that is what they did to Earth. The first settlers knew this. They knew that as soon as the chemistry of the atmosphere had been changed, and temperatures had increased sufficiently to support their plant crops, they could land, and, having landed, they had to begin at once to cultivate as large an area as they possibly could. They needed to make the planet live, in order that its life might make it more habitable. As you know, their success was spectacular, but that is because they understood very well what they were trying to achieve.

The realization that living organisms can and do modify their environment to suit themselves came slowly, for it seemed to some people that the idea contained a fatal flaw. The theory of natural selection as the principal mechanism for evolution had become something of a dogma among biologists. This was unfair to Darwin, who first propounded the theory, for he was anything but dogmatic and would have welcomed new information that might have modified his ideas. As it was understood by Darwinians, the theory proposed that life forms adapt to the conditions in which they must live, the conditions themselves being supplied by the inanimate processes of physics and chemistry. It, or more properly they, had no room for the proposition that it is the organisms that determine the environmental conditions. In due course they yielded and the concept was adjusted to take account of the new knowledge.

The more serious objection, though, was held to be the teleological element in the new theory. It looked as though organisms had some means by which they could predict the conditions they would prefer, and then worked together to attain them. In other words, the present was being conditioned by the future, cause and effect were running backwards, and one of the major scientific heresies was being propounded.

I think I mentioned earlier the way in which such *a posteriori* reasoning had been used by some people to 'prove' that Earth must be special in some way for anything so complex as living organisms to have appeared on it as the result of 'chance'. The words 'chance' and 'randomness' seem to cause difficulty. If I throw a pair of dice the resulting score is a matter of chance, and the upper faces of the dice are chosen at random. That is to say, I am quite unable to predict what it will be. If I throw the dice many times, however, I can predict to within fairly narrow limits the number of times each score will appear. So in this sense, although each event in the series is produced by a random process, the randomness disappears when many events are considered. Evolution by natural selection involves an element of chance, in that we cannot predict each of its events because they occur at random. However, they also occur, as gene mutations, very frequently, and their success, their ability to survive and establish themselves, depends on the qualities they confer on the organisms possessing them. Here, in theory at least, we can make predictions. Thus, the material on which natural selection operates is generated randomly, but the process of natural selection is not random.

It was all a misunderstanding, for the new theory was not in the least teleological and did not require that organisms exercise free choice, far less predict the outcome of their lives. The simplest way to explain how a living planet works is to use the example that was devised back in the early 1980s, when it was called 'daisy world'. Let us talk about daisies—now we have erected our 'greenhouse'.

I ask you to imagine a planet where the temperature is sufficient for plants to grow. However, the only plants that do grow are daisies, and some of them are white and others are black. Lacking all competition, they form dense stands. Between themselves they are as aggressive as species can be, and the disadvantages of one type of daisy will be exploited by the other. On Mars, the black algae were the equivalent of the black daisies, and the pale martian soil the equivalent of the white daisies.

Over the years, the temperature of the planet will tend to change. Sometimes it may grow warmer, sometimes it will be cooler. How does this affect the daisies?

The warmer the conditions, the faster daisies will grow. However, if it becomes too warm or too cool, growth will be slowed and finally it will cease.

Let us suppose the planet receives less radiation from its star, so that the climate tends to grow cooler. The black daisies, having a lower albedo, absorb more radiation, so the temperature inside their stands is higher than the temperature away from the stands, and markedly higher than the temperature among the stands of white daisies. Therefore, being warmer, the black daisies will grow faster than the white daisies, and will invade areas formerly occupied by white daisies.

If the radiation increases, and the climate tends to become warmer, the temperature among the stands of black daisies will still be higher than the temperature among the white daisies, but now it may begin to approach levels at which growth is slowed. The white daisies will grow faster, and this time they will invade formerly black areas.

As you see, the daisies themselves only respond to the conditions they experience. They are not required to make deliberate choices, far less to predict future conditions. They merely react, in a most unteleological way.

Their reaction, which is an effect following a cause, is also a cause of other effects, however. As the area occupied by black or white daisies increases, the albedo of that area changes. In our imaginary world, daisies cover the entire land surface, and so they have a very large effect on the albedo of the planet itself. Let us look again at the sequence of events.

The amount of radiation increases. This causes an expansion of the area occupied by white daisies, which in turn increases the planetary albedo. The planet is now reflecting more incoming radiation than it was, and so the surface of the planet is being warmed less than it would be were the daisies not there. Indeed, if the daisies are very white and very black, the temperature of the soil beneath the daisies will change hardly at all. If the amount of radiation decreases, the black daisies will expand, and again the temperature of the soil will be held fairly constant.

Let us allow that the air is free from solid particles, which can absorb or reflect radiation and so confuse the picture. In this case, the air is clean and so it will be warmed by contact with the ground. The climate is determined by the temperature of air masses. As the amount of radiation received by the planet increases or decreases, so the proportions of white and black daisies alter in response, and the ground temperature, and the temperature of the daisies themselves, remains fairly constant. Therefore, the temperature of the air, warmed by contact with the ground, also remains fairly constant. This means that although the amount of radiation may alter, the response of the daisies has the secondary effect of maintaining a more or less constant air temperature, and thus of minimizing climatic change. In fact, the daisies are controlling the climate, preventing it from becoming too warm or too cool for their growth, and they are doing so quite automatically. They are not conscious, they predict nothing, plan nothing, anticipate no outcome. They simply grow and compete among themselves.

The theory sounds plausible enough, but does it work? Prior to the Mars landings, it had not been tested on a real planet, but it had been examined rigorously. Numeric values had been given to all the variables involved, they had been fed into computer models, and the models produced descriptions of theoretical worlds on which the climate was controlled by the daisies. The settlers knew of this, and it was they who conducted the first 'field trial', on a planetary scale.

Perhaps I have made it sound simpler than it seemed to those involved. The 'daisy world' approach is what was used, it worked, and we have the wisdom of hindsight. At the time, when the first voyages were being planned, 'daisy world' existed only in theory. Many scientists believed it should work, but it had not been tested. It could not be tested on Earth, for on Earth the living organisms had been in control for several billion years and the only test possible would have involved disrupting their system to see what would happen without them, which was hardly practicable!

Meanwhile, there were other possible approaches that had been tested. The chemical industries, for example, had many years' experience of producing the gases and other materials the settlers would need. If Mars were turned into a huge chemical factory, using technologies developed and perfected on Earth, that might have proved simpler and to many people it sounded much more reliable.

The settlers were going to need vast amounts of substances that were taken for granted on Earth. Until then, no one had ever had to stop and consider how to supply oxygen to the air, or nitrogen, or phosphorous and potassium to the soil, how to find calcium, sodium, or any of the elements that are essential to terran organisms. They were proposing to build a whole environment, starting almost from scratch. What is remarkable is not that they paused, and argued, but that they agreed so rapidly on a solution that proved so successful.

Fields or Factories

The decision to construct a 'daisy world' on Mars was brave, for by the time the attempt was made many people had staked their reputations and in some cases their life savings—not to mention their lives—on its success. It was especially brave because there was an alternative.

This ship, in which we are sailing the skies in safety and comparative comfort, provides the clue to that alternative. By the late twentieth century human engineers had achieved great things. Men had walked, indeed driven, on the Moon. An unmanned vehicle had explored the outermost planets of the solar system and had left that system entirely, still functioning. Factories, manufacturing complex goods that had joined the list of necessities for life in an industrial society, had ceased to be the filthy, unhealthy workplaces they had been a mere half-century earlier. They were clean, and the repetitive, unpleasant, dangerous work was performed by robots. There were unmanned trains to transport human passengers about great cities, and young children played expertly with small, inexpensive computers with a computing power that only a decade or two earlier would have required a machine the size of a large room and would have cost at least a thousand times more. Humans had been connected directly to computers, so that where great skill, judgement and precision were needed in the performance of complex operations, the human nervous system supplied instructions directly and immediately to electronic symbionts. In every area of engineering and technological skill, progress was being made so rapidly that there seemed to be no limit to human ingenuity. Consider the ships that were being planned for the journey, that would move as tiny, habitable oases, carrying with them all that humans need to live, and think of the engineering accomplishment they represented.

Mars could have been rebuilt, physically and chemically, as a replica of Earth, rather than being encouraged to develop in its own ways. All the technical skills needed to restructure it existed, and most of them had been tested.

If the technocrats had remained silent, if the opportunities they offered had been unknown, the choice might have been easier. People who were not engineers or scientists might have speculated, but they could not have known what was really attainable. As it was, the technocrats were anything but silent. They were as keen as anyone to encourage space exploration, and they drew up plans for the colonization of planets that extrapolated from their knowledge of space vehicles. They advanced their plans, they lobbied for funds, they sought publicity in order to bring popular opinion to bear on politicians, and the 'package' they proposed was well known. No one with an

interest in space exploration was allowed to remain igno-
rant of it.

Their plans for Mars began where all plans for Mars had
to begin. The environment on the planet had to be altered
to make it habitable for humans. To do this, atmospheric
temperatures and pressures had to be increased, the chemical composition of the atmosphere had to be modified,
eventually to make it breathable, and necessary chemicals
and minerals had to be obtained from martian rocks and
soil.

The temperature of the planetary surface might be increased in two ways. Large mirrors might be parked in
orbit about Mars to reflect the sunlight on to the ground
below. This would increase the amount of solar energy the
planet received. The mirrors themselves would be very
large, each of them hundreds of square kilometres in area,
but they would be made from thin material that weighed
little, and they might have been no more difficult to
manufacture and set in position than the large sails of the
early space sailing ships. At least, the technocrats sounded
very confident that they could manage such unwieldy
items.

In conjunction with this, the planetary albedo would
need to be reduced. The surface would have to be darkened.
Both groups, those who favoured this technocratic approach and those who preferred the more horticultural,
'daisy world' approach, agreed about this. The technocrats,
however, would have darkened the planet by using dust, or
finely comminuted rock, which they would have obtained
from Phobos and Deimos, the two martian satellites, both
of which are dark in colour. Human workers would live
deep inside the 'moons', controlling the collection and
discharge of material. The dust would be aimed primarily
at the two polar icecaps and at areas of known permafrost.

Finally, in order to maintain the warmer conditions and to accelerate the melting of the permafrost, energy could be released at the martian surface by means of a contrived planetismal impact. You will remember that I have mentioned such impacts before. They have occurred many times in the past, on Mars and on Earth, too, and the arrival of a body perhaps ten or more kilometres (six miles) in diameter, travelling at some 72,000 kilometres (45,000 miles) an hour releases a great deal of energy, almost all of it as heat. In this case one or more asteroids would be captured and thrown, perhaps by means of large explosions in their vicinity, into orbits that would carry them away from Mars, around Jupiter whose large gravitational pull would exert a kind of slingshot effect, and so back at Mars again at very high speed.

It might have been feasible. All of it might have been feasible, and what is more every last item in the scheme could have been calculated very precisely, using known laws of physics. It called for no new materials, no technology that did not exist already, and it might not even have been prohibitively expensive. Some hundreds of square kilometres of sheeting made from some aluminum alloy, propulsion systems, a few thermonuclear bombs that could be better employed in the asteroid belt than on Earth, and the means of transportation for humans and their tools to another part of the solar system, were all that would have been needed.

Once they arrived, the first humans would begin to exploit the changes that had been made. They would find that in the deepest hollows, those made recently by asteroid impact, it would be possible to plant terran crops. They would build shelters for themselves, mainly by cannibalizing the ships that had conveyed them, and they would bury their shelters beneath the martian soil for insulation and for protection against cosmic radiation. They would walk in the open wearing spacesuits, and would maintain their buildings at terran atmospheric pressures.

Yet another technocratic approach might have been attempted. Humans might have landed on Mars without any advance preparation of the planet to receive them, as the first humans landed on the Moon. They would construct shelters using parts from the ships that brought them, and then they would start building factories, real factories like those on Earth. The factories would use raw materials obtained from the martian soil to manufacture the gases that were needed to produce a 'greenhouse', as well as the oxygen and water needed by the colonists. The chemical composition of martian soil was known, and the chemical technologies required were also known or could be devised and tested on Earth before any human life was staked on them.

There were many people who found one or other technocratic approach attractive. The engineering seemed reliable because it required no new knowledge and, indeed, it would be a mistake to make too much of the differences between the 'engineers' and the 'biologists' for this was not so great as it might seem. True, in the end terran plants and micro-organisms were used to modify the martian climate, but humans still had to build factories, still had to supply themselves with necessities their non-human colleagues could not provide. The question was never whether factories should be built or not. It was less than that, yet more fundamental. It centred on the question of whether Mars should be made into a replica of Earth, or whether it should be 'seeded' with life, and allowed to develop in its own way in its own time. The dispute concerned the goals that were sought, rather than the means by which those goals might be attained.

When I remember what I was taught as a child, when I recall what I have seen of the terran newsreels and recorded discussions of those times, it becomes clear just how difficult that initial choice was. Lives were to be staked, after all, on a theory that had not been tested. Yet there were several factors that had to be considered. In the first place, human attitudes to Earth had altered radically during the 1970s and 1980s. The planet was still highly industrialized, of course, but more subtly than it had been. Biotechnologies were employed in tasks, such as mineral extraction, that formerly would have required large machines and would have devastated landscapes. The conservation of species and their habitats was accorded high priority and any proposal to interfere with a natural plant and animal community was examined very critically. The days were long gone when a company could simply erect any kind of factory on any site that seemed suitable. The idea of 'greening' Mars, rather than 'industrializing' it was inherently attractive. It seemed gentler, more 'natural', and it was more natural too, in that the transformation was to be achieved by the activities of living organisms, left to their own devices.

There was a second consideration, and now that we are so far on our way from Earth I dare say this will carry more weight with you than it might have done before we departed. If Mars were to be turned into a kind of factory, all the processes would have to be designed on Earth. The industrial plant would have to be made on Earth and assembled on Mars. What would happen if something went wrong after the humans had arrived? Obviously they would be in radio contact with Earth and could ask for advice. But what if a serious breakdown could be remedied only by the replacing of a component and the replacement had to be sent from Earth? Would it arrive in time?

What if the instructions sent to Mars could not be implemented because of some local circumstance that had not been foreseen? People began to brood on such matters, and the image of a planet-wide factory began to seem less persuasive.

I think the final thought, the one that tipped the balance in the minds of the first colonists, was more political. People who migrate to pioneer new lives in new lands are usually of an independent turn of mind. Terran history shows that many of them quarrelled with the authorities in the countries they left. They liked, indeed demanded, a freedom from restrictions imposed by others. Had our pioneers taken with them to Mars kits for making do-it-yourself factories they would have been heavily indebted to terran industries. Probably representatives of the firms involved would have had to travel to Mars to supervise much of the work. Those representatives would have continued to be employees of the firms that sent them. They would have reported to a terran authority, would have acted on the instructions of that terran authority, and on Mars they would have had immense political power, backed by the threat to withdraw the support without which the enterprise would fail.

If industrial organizations were to possess power of this kind, terran governments would insist on supervising them. So the final authority would rest with the governments of Earth. Such political control would be justified on two grounds. The security of the migrants required control of the industries on which they relied, and the investment cost of the operation would need to be underwritten by governments, who would then remain on hand to safe-guard public money. It was the kind of arrangement that was unpalatable to the pioneers, and if life and death decisions on Mars were to be taken ultimately by politicians far away on Earth, almost certainly it was highly dangerous. A better way had to be found.

Indeed, there were examples to hand on Earth of the long-term consequences of placing the control of territory in the hands of a private company. Groups of islands had been allowed to develop in this way and their human societies had atrophied, the opinions and attitudes of their people frozen into a complacency that refused to acknowledge changes that took place in the world beyond their small patch of ocean. Theirs was a way of life that was incapable of adapting to change, and survival depends on flexibility.

The 'greening' of Mars was highly dependent on industry, but it was dependent in a different way. The initial transformation relied on natural organisms, transported by the Foxe rockets that had been more or less donated, and industrial assistance was needed only after the first colonists had arrived and established themselves. The dependence on industry was much the same, in fact some ideas were 'borrowed' from the more technocratic prospectuses and used with great success, but the industry itself would be based wholly on Mars, not on Earth. The factories would be built on the spot, from martian materials. There might be industrialists, but they would be martian industrialists—the colonists themselves. From the first day, indeed from the time the first humans left Earth, their fate would be entirely in their own hands.

You can see Mars very clearly now. The screens that used to display various views of the sky now show only that part which contains our destination, with Mars by far the brightest object apart from the Sun itself, exactly at the centre of the picture. All the screens, in all parts of the ship, show the same view when they are not being used for entertainment.

I notice it especially at night, when lights are dimmed on board and most people are asleep. Sometimes I walk a little at those times, and recently I have found others, the insomniacs I suppose, sitting or standing and just gazing at a screen. They do not talk. Each is locked away tight inside his or her own head. The time is near when the dreams must be tested against reality. It is sobering and the passengers are becoming nervous.

How should I reassure them? It seems pointless to explain that Martians are only people, after all, people like themselves and understanding of their feelings. They will be welcomed on Mars. They know that, or should, for they have been told it often enough. It is their imaginations that create the fears, and the fears are of themselves. It is a big step they have taken, and perhaps it is of little value to assure them that many, many before them have taken that same step.

For what it is worth, I can tell of how much more frightening it must have been for the first humans to arrive. They were prepared, of course, as well as they could be, but for what? I have said before and I repeat at every opportunity, their situation was quite unlike that of the first explorers in America or Australia.

It was not simply that the new habitat differed. The people differed, too. In the first place there were more of them and their organization was much better. The need for organization is obvious, but it was also necessary to move a fairly large number of people to provide a kind of 'critical mass' of labour. Part of their preparation involved the planning of the work that would have to be completed according to a strict schedule. There was to be a great deal of work, and it needed many hands. In time all manual work would be performed by robots, of course, but in the early days there was neither space on the ships to transport robots, nor time on Mars to build them. They had to come later. The first settlements had to be built by humans. Perhaps it will be best if I explain in a little more detail just what that work entailed.

As I said, there was a schedule that had to be observed precisely. The initial colonization had been planned in such a way that groups of people would arrive, each group joining those already there in preparing for the next group. The arrival of each group was timed back on Earth, of course, because it was determined by the date of departure, so that those on Mars had to complete their tasks according to a timing over which they had no control, and the colony had to be established before the onset of winter. Winters were much less severe than they had been, of course, but it was necessary to have plant crops established and growing before growth slowed as temperatures fell.

The key to the operation lay in the fact that the ships by which people travelled to Mars would not return to Earth. It was a one-way trip for both ships and people. Once they reached Mars the materials from which the ships were made would be used—and the humans would remain. They could not change their minds.

As soon as the first ship was parked in its orbit of Mars a lander left it for the surface, carrying ten passengers of whom one was a qualified astronaut, some tools and supplies, and enough fuel for it to return to the ship. The first arrivals unpacked the lander and built a temporary shelter that would accommodate the full complement of fifty people who had undertaken the journey.

Meanwhile, those remaining in orbit began to dismantle part of the interior of the ship itself. The surface shelter constructed, the astronaut returned the lander to the ship for the next cargo, of nine more passengers (ten counting himself), and supplies which now included some that had been part of the ship. The supplies were unloaded and stored in a crude stone lean-to shelter that had been built while the lander was collecting them. From that point on, the remaining people on the ship donned spacesuits, and the lander carried down cargo that consisted increasingly of the ship itself. One of the last loads consisted of the fuel rods for the plutonium reactor.

The final load was different. The lander was parked in orbit and the remaining humans tethered themselves to it. A part of the outer skin of the ship was fitted together to form a casing around the reactor. During the journey to Mars the reactor had not been used and it contained no highly radioactive waste material. Power for the systems on the ship had been supplied by fuel cells, augmented by solar cells, and the reactor had been held in reserve.

Between the casing and the reactor, a fairly large space was packed tightly with a fine but dense 'wool' made from carbon fibre. In the vacuum of space, heat can be lost from the skin of a ship only by radiation, and this packing had provided thermal insulation between the inner and outer skins on the ship. It was far from being the most efficient insulating system that could have been installed, but it was the only one that would work and that would also supply necessary packaging material on arrival.

Small auxiliary chemical rockets were attached to the outside of this casing, the rockets were fired retroactively, and the reactor began its descent while the humans returned to the lander and waited until it had arrived.

I should point out that in some ways an entry of this kind is a far less hazardous undertaking on Mars than it is on Earth. The much thinner atmosphere slows an object more gently, and the fact that the martian atmosphere extends much further from the planetary surface than does the terran atmosphere means that the braking effect continues for much longer. Friction is much less, and so the heating against which Earth-orbiting vehicles need highly efficient protection can be absorbed by only a modest shielding made from any of a wide range of common materials. The terminal velocity of objects falling through the martian atmosphere is not much different from that on Earth, because the lower atmospheric friction and the lower gravitational pull cancel one another.

However, parachutes cannot be used in the martian atmosphere. Even on Earth, aviators are advised never to attempt to open a parachute above an altitude of 3000 metres (10,000 feet) because the air density is so low that it may fail to open properly and may tangle in its own rigging. On Mars the atmosphere is much thinner than this, even at the surface. So a parachute descent is not recommended. Instead, the reactor package was allowed to fall freely, all the way to the ground. The careful packaging protected it from damage. It had all been calculated with great care, and those involved knew what they were doing. In fact all the calculations they needed had been made on Earth long before, in the early days of martian exploration, when plans had been made simply to drop instrument packages on to the surface. With the packaging provided for them even sensitive instruments would have survived the impact. The plans were shelved for those unmanned explorations and soft landings were made instead, but the calculations remained valid. It was necessary only to apply new values to the equations.

The procedure is simple, but it is not very accurate. By timing the firing of the rockets the point of entry can be determined well enough, but in the course of its descent an object will traverse a large area. Thus, although it was possible so to arrange matters that an object would land on smooth ground, and somewhere not too far from the place it was supposed to be, once it was down it had to be found and then moved to the place where it was needed. The lander was used for this purpose, and that is why it remained in orbit until the final load had landed. Its crew located the reactor, arranged their own descent to place them close to it, and then used the lander to carry it to the site of the settlement.

Not quite everything from the ship was taken to the ground. The sails would have been of little use on the surface. They were very fragile, about as thick as gold leaf, and any attempt to fold them and then open them again would have damaged them. If erected on the surface, assuming that would have been possible anyway, they would have been destroyed by the first martian breeze to strike them. They were detached from the ship, turned, and parked in a stationary orbit so they reflected sunlight on to the ground in the area of the settlement. This was one idea taken from earlier proposals. It differed in only one respect, albeit an important one. The 'mirrors' were not sent to Mars separately, and in advance of the landings. They were used all the way to Mars, which 'paid their fare', then left behind usefully. The work they did was not part of the initial 'greening' but was intended only to continue it.

The materials transported to the surface were used to build the first permanent shelter, which was completed within a few days. The reactor was charged with its fuel and began to provide power, and life on Mars began.

Once that shelter had been built and the reactor had been started, the first stage was complete and the colony was ready to welcome the next arrivals. They came in the same way, bringing their ship with them and using its materials as well as its cargo to add to what had been built earlier. So it continued, with party after party landing and expanding on the work. Eventually the first colony, the people who spent that first winter on Mars, was one thousand strong. Later I will explain in more detail who they were, how they were chosen, and how the operation was financed. For the moment I will concentrate on what they achieved.

A large conservatory was built to grow food by a very simple hydroponic technique. Semicylindrical sections of alkathene guttering were erected in parallel on supports so that they sloped slightly from an upper to a lower end. At the lower end they were linked by more guttering which also sloped and fed into a small electric pump. This was connected by piping to a reservoir, and from there to a feeder pipe at the upper ends of the gutters. When the reservoir was filled with an aqueous solution containing plant nutrients and the pump switched on, a thin trickle of nutrient solution flowed down each gutter, into the gutter at the bottom and so to the pump and back to the reservoir. Seedlings, raised in trays, were transplanted directly into the gutters, where they could be crowded together. The use of nutrient and water was frugal, and yields were huge.

It might have been possible to produce food simply by culturing plant tissue and not growing whole plants at all. This was considered, but rejected for fear that it might prove unreliable on a large scale. There were those who maintained that the best machine for growing plant tissue is a plant! Animal tissue might also have been cultured, but this scheme was dropped for the same reason, and because animal products were unnecessary to the success of the enterprise. People felt very strongly that the simpler the technologies they used, the greater would be their chances of success.

This intense plant growth soon began to assist in the

recycling of air, that to start with had relied on the system brought down from the ship, which indeed was the source of the air itself inside the building. The plants were not an important source of oxygen in the air, however. While they were growing vigorously, and incorporating carbon into their tissues, there was a net oxygen gain, but it was too small and too unreliable to be taken into account.

The provision of oxygen was much simpler than that. The martian soil was rich in pernitrates, highly oxidized substances that liberated oxygen when they were warmed. In this, Mars differed greatly from Earth. When life first appeared on Earth, there were large amounts of ferrous iron oxide (whose molecules comprise one atom of iron and two of oxygen). As photosynthesizing plants released oxygen, so the ferrous iron was oxidized further to the ferric form (one iron atom to three oxygen atoms). Thus the terran surface was reducing, it 'scavenged' oxygen and held it securely. The martian surface was oxidizing, and released oxygen very readily. In fact, it was necessary only to collect soil and store it at ordinary room temperature for oxygen to be liberated. An early task was the preparation for the building of a factory that would separate martian ores from their oxides, because the metals were needed, and this process yielded oxygen as a valuable by-product. The first party did not build the factory, of course. That took some time and was not completed until late in the first season, although it was working by winter.

The ease with which modest amounts of oxygen can be obtained has saved a few lives. All martian children are taught that if their breathing apparatus should fail while they are in the open, they are to spread their outer garments on the ground, lie face down beneath them, and the warmth beneath the covering will produce a little oxygen. When breathing equipment fails it begins to emit an emergency radio signal, and the oxygen from the soil can be just about enough to keep a person alive until they are rescued.

Some ground near the settlement was covered with a transparent plastic, so that as water was melted by the orbiting mirrors it could be captured before it evaporated. The water was purified, for it contained high concentrations of mineral salts, and much of it was then used to wash similar salts from the land on which outdoor plants were to be grown. Martian soil was intrinsically suitable for terran plants, but repeated melting and evaporation of water from it had concentrated salts in its upper layers. Such salination was a common problem on Earth, then as it is now, and its remedy was well known. It had not caused problems for the algae and bacteria, which included species that were tolerant of highly saline environments.

Those first organisms included some that had been modified genetically to make them capture and concentrate particular plant nutrient elements, so that by the time the first humans arrived, the soil was already becoming fertile. Again, the technology for making such genetic modifications had been developed on Earth long before, and microbial mining for valuable metals was established back in the 1970s.

As soon as a site had been prepared, sowing began. It was vital that the 'greening' of the planet proceed as rapidly as possible. The 'crops' the first Martians sowed were not necessarily edible by humans. Their purpose was to transform the planet, to build on the process that had been started by the algae and bacteria, not to feed its human inhabitants.

Again, the technique was well established and simple. Seed, comprising a mixture of hardy, moderately salt-tolerant, drought-resistant plants, was mixed in water with the organic conditioner made from human faeces, mainly during the journey in space. On Mars, the plants were taken partly from desert environments on Earth, but they also included many seeds and spores of ancient plants, relics of the time when Earth had less oxygen in its atmosphere than it has today, such as horsetails. These are mostly fairly small, herb-sized plants that prefer shade, but there are some that grow much larger, to twice the height of a human, so from the start Mars began to produce 'horsetail woodlands'. A pump was used to spray the mixture from a jet, so that large areas could be seeded very quickly. The seeds germinated rapidly and were held in position by the organic mulch which dried around them. The nutrient contained in the mixture provided the seedlings with sufficient nourishment to allow them to become established, and once they were established, with the help of the microbial miners' who were working all the time, the process became self-perpetuating and the fertility of the soil improved rapidly.

Not all areas could be colonized in this way by the larger plants. Much of the higher latitude regions of Mars were, and still are, dry desert. However, even dry desert is not devoid of life. The larger plants could not live there, but the algae and some lichen could, and did, contributing to the darkening of the surface.

Even those algae and bacteria that had not been altered genetically had been at work since ever they began arriving. They had been darkening the surface, but they had been doing much more than that. They had been extracting from the martian rock and dust the nutrients they needed, which included many of the trace elements that other plants need in minute amounts. They had been dying and reproducing, and so they had begun the process of soil formation. Yet again, it is what happens on Earth, when apparently bare rock, or a wall built by humans, is colonized by species that dissolve nutrients from exposed surfaces of the rock itself and so commence the process of soil formation. The seeds sown by the first colonists derived much benefit from the preparations made by the algae.

One of the major contributions made by the environmental movement on Earth had unwittingly exposed the fallacy in an idea that had been held for many years. The environmentalists had drawn attention to the large areas of wasteland left behind by industries when their operations ceased and had demanded that such land be reclaimed. In the course of that reclamation knowledge was gained that could be applied on Mars once it was realized that Mars was rather similar to terran wasteland.

The fallacious view, repeated in many textbooks, was that a soil suitable for the growing of large plants takes centuries or even millennia to develop. Figures were quoted, suggesting rates of soil formation measured in centimetres per century. No doubt this is true where soils form by processes in which humans do not intervene. When humans intervene, however, it can be accelerated dramatically. Industrial wastes, comprising ground that was compacted, acid, deficient in nutrients of all kinds, were reclaimed using hydroseeding techniques just like those used on Mars, and soil was made to form at a rate of centimetres a year. The mistake had been to suppose that large plants, such as grasses, cannot grow except in a soil rich in organic matter. Provided they can be anchored securely and supplied with the water, nutrient, and sunlight they need, they will grow almost anywhere. They do not need a rich topsoil at all. Soil formation was very rapid, and after no more than a year or two could be established and farm crops produced.

Yet again, I fear I have made it all sound simpler than it was. If they are to support plants, soils—even apparently inert ones—require more than manure and seeds. They contain micro-organisms in vast numbers, and soil animals, every one of which has an essential role to play in the recycling of organic wastes and the chemical processing of mineral nutrients. This vast and complex population had to be supplied. The algae and bacteria that were already established had adapted to the conditions they found, but they were not enough. Fungi were needed, and protozoans, nematodes and mites, tiny spiders and insects, and worms. All of them had to be carried from Earth.

Had it been necessary to work out in precise detail the composition of the population that was needed, those pioneers might have been on Earth yet, still earnestly compiling their inventories. In practice it was very simple indeed. They took with them samples of fertile soils, complete with the populations they contained. They had soils of various types, and they had many different samples, kept isolated from one another for safety, and these were kept inside their buildings and tended carefully until the time came for them to be 'sown'. Once the plants were established and dead vegetable matter was beginning to accumulate, the imported soil was packed thinly on to the ground and mixed with martian soil, much as compost might be applied to a garden. With a large supply of food and no competition the decomposers multiplied rapidly and it was not long before martian topsoil, though thin, came to resemble terran topsoil.

By the commencement of the first winter, then, the first martian colony was established. It had its own shelter, its own food supply, its own air. The sewage it produced was used to supply nutrients for plants, some of it indoors, but most as the mulch used in hydroseeding which continued until the growing season ended.

From that time, the further colonization of the planet was assured. As the second spring began and the next batches of humans began to arrive, they were taken into the established colony temporarily and used it as a base while they repeated the construction, this time with the benefit of the experience of the old hands, and so made their own settlement nearby. That pattern was repeated many times, each new group using an established colony as a base from which to build their own settlement, and little by little people moved further away from the original site.

Problems arose, of course, and there were tragedies. One of the early settlements, built in the fourth martian year if my memory serves me correctly, was destroyed completely. They had misjudged the siting of their buildings in relation to the orbiting mirrors. This was a delicate business for everyone in those days, and danger could creep up unannounced.

I should explain that martian skies are not clear. Most of the time we have a thin haze, rather like terran photochemical smog. This is due to the chemistry of our atmosphere and it was not planned. It results from the course evolution decided to take on our planet. The growing plants remove carbon dioxide from the air, as on Earth. When the plants die, however, their decomposition works differently. The organisms most involved in it return the carbon from the plants to the air not as carbon dioxide, but as methane. The methane-producing oranisms were brought from Earth, of course, and methane is produced on Earth in this way, but on Mars, where the air contains little oxygen to support the oxidation of carbon, they proliferated. In the air, the methane is oxidized to carbon dioxide and water, but the complex reactions involved—for the reaction proceeds in several stages—produce the haze.

The point is that the direct sunlight reflected by the orbiting mirrors does not produce brightly lit patches of ground surrounded by more dimly lit areas. It is not very easy to see at all, and shades almost imperceptibly from the warmer to cooler ground. The warmth melts the permafrost, very slowly. When permafrost thaws it does not release liquid water at once, to form rivers or marshes. Long before that stage is reached the soil itself is 'loosened.' The ice that bound soil particles together becomes liquid, particles begin to slide past one another, and on even quite gentle slopes the soil itself begins to flow. The process is called solifluxion and it is well known on Earth, but on Mars the heating from the mirrors made it a very local phenomenon in the early years.

It meant that there were two kinds of site on which permanent buildings must never be constructed. Unless their seismic surveys made them absolutely certain they were above solid bedrock, people must never try to live inside the area being warmed by the mirrors, and they must never live at the foot of a slope below an area that is being warmed. Indeed, it was best not to live at the foot of a slope at all. The second kind of site was easy to avoid, because slopes can be seen clearly. The first site was not difficult to identify, but where the ground felt and looked firm people could be careless, and my ancestors were fortunate in that disaster struck only once.

In this case the permanent buildings were erected on ground that was being warmed but that was not so secure as it seemed. The ground sloped—most ground does—but for a time the weight of the building compacted the soil beneath it sufficiently to hold it in place. Unknown to the occupants, it was being undermined several metres below the surface by soil and water flowing downhill, and when at last it gave way it did so catastrophically. No one survived, and so our knowledge is based on what was pieced together later from the evidence gathered by the rescue team. It must have been much like an earthquake. Suddenly, without warning, the entire building collapsed as the ground beneath it moved. Like most martian buildings it had been partly buried, so that the upper part was secured by dry soil the settlers had put there themselves. The upper part tried to remain in position, while the lower part moved with the ground. Some people were crushed, others asphyxiated. It was appalling, and from that time people were a great deal more careful.

Descriptions of the unfamiliar raise more questions than they answer and my reminiscences are patchy at best. I remember what I can, describe what I know, but when I watch the anxious faces around me it all seems woefully inadequate.

Yet I continue, and perhaps I help. I realize that I must explain how the settlements arranged themselves geographically, because that explains much of the way in which our society works. I have mentioned a rescue team, but I have not said how it travelled across martian terrain. I must do that. At the same time I must explain the very existence of a rescue team at all, because in doing so I can demonstrate that our colonization followed a different path from the colonization of Earth.

Before I can embark on these explanations, I must keep the promise I made earlier, to say something about those early settlers. I owe it to myself as much as to my fellow passengers, to affirm that I am descended from neither convicts nor political subversives. My ancestors were not even 'cowboys'. It was all rather dull, I suppose, but highly respectable. I must explain, too, how those early trips were financed.

That is where I shall begin, for it is a story in itself.

The First
Footprints

On Mars we do not eat animals, or use any product
derived from animals. Some of my terran friends are
distinctly carnivorous in their tastes and have complained
about the quality of food on the ship, of martian food. I
reflected some time ago about the number of complaints
there had been about the cuisine, and was privately amused
by the vagueness of the replies such complaints receive.
There is not much the crew can say. They know, and I
know, that no Martian would complain.

Terrans eat meat only from habit. In their early history it
made a great deal of sense. You can obtain highly nutritious food by killing animals, or by stealing prey from
other carnivores. Then, later, some of the prey species
were domesticated, much of the land was suitable only for
converting to pasture, and so livestock farming became
established rather firmly, and any suggestion that it might

161

be abandoned, and the livestock farming regions be turned into nature reserves or parks, was resisted strenuously and not only by the farmers. Yet animal farming ceased to be necessary before the end of the twentieth century.

Animal products are highly nutritious, and it may make sense to keep livestock on land for which no other use can be found, but it is very inefficient. If you feed plant foods to animals, then eat the meat yourself, you obtain no more than about one-tenth of the food value of the original plants. If you could find a way to eat the plants yourself directly, not only could you dispense with the animals, but you could feed yourself from a much smaller land area.

It became possible to do that when ways were found of extracting nutrients, and especially proteins, from plants that were inedible for humans. Plant cells are encased in walls of cellulose, and mammals cannot digest cellulose. Ruminants employ large colonies of bacteria to do the job for them, while horses eat vast quantities of food, and pass it through their guts rapidly, extracting a tiny fraction of its food value as it passes. The human digestive system is intermediate between that of a true carnivore and that of a true herbivore, so that we can eat most kinds of meat and some plants. Our ancestors developed leafy plant varieties whose cell walls were thin, and fed on storage organs, such as fruits, seeds, tubers, and roots, that they could digest.

Once it became possible to break down the cellulose industrially, grass became edible even for humans. They invented 'plant milk', a milk-like food that was made from plants but without the assistance of cows; they extracted proteins from soya, wheat, and other protein-rich plant products, and they grew cultures of yeasts on carbohydrates, and harvested them to extract protein, so converting carbohydrates that are easy and cheap to produce into proteins that are much more costly. What did they do with most of these novel foods? Why, they fed them to their livestock and went right on eating meat. Terran behaviour can be difficult to understand sometimes.

Animals were also used to obtain other products. At one time this was very important. Furs, skins, tanned hides, bones, and wool supplied necessities that were difficult to obtain from plants, especially for people living in cool climates, but alternatives for all such products had been found by the middle of the twentieth century.

There were vegetarians on Earth in those days. Their numbers have increased very substantially since the twentieth century, but most of them avoid animal products either because they do not like the taste or texture of them, or because they feel it is morally wrong to kill animals for food when there is no need to do so. Martians would agree with the moral viewpoint, I imagine, but that is not the real reason for making the whole of our population vegetarian. It is much more practical than that.

Air-breathing animals cannot live outdoors on Mars. They can live in the soil, where oxygen becomes trapped among soil particles, but they cannot walk around in the open. So our wildlife contains no reptiles, no birds, no mammals, no large insects. We cannot grow brightly coloured or scented flowers outdoors. There is no meadow on Mars to delight the summer eye and nose. Such flowers rely on animals to pollinate them, and on Mars the strategem cannot work.

The first glimpse of the martian countryside can be a shock. The rolling hills are dark green beneath a pink sky, and in the mountainous regions the snow caps glisten throughout the year. We have horsetail woods, a few larger trees, and even some small forests, but the martian green is not relieved by the blues, yellows, whites and scarlets of flowers, for while we have some flowering plants their

flowers are wind-pollinated and usually green. The green is
relieved. We do have patches of bright colour, supplied for us
by our algae and lichen as splashes of orange, dark red, and
purple, but we like green as a colour. The greening of a new
area as plants colonize it is greeted with enthusiasm, as a sure
sign that the planet is evolving rapidly—but to what? Perhaps
it is evolving toward a condition in which air-breathers can
walk and fly, flowers can grow, and an abundance of oxygen
will give us a terran-blue sky. Even then, you need a lot of
oxgen in the atmosphere, and a much denser atmosphere than
we have, before animals can fly. The day of the bright carpet
of flowers may never dawn for us. Earth is a riot of colours.
Some might call it garish, vulgar. Green is good!

I might mention that it is only very recently that Earth
acquired brightly coloured flowers. Humans think they
have always been there, because they arrived before the
humans did, but they have not existed for more than about
a hundred million years.

The point is that if we wish to keep animals, as 'slaves'
or as pets, we must do so indoors. That would be possible,
of course, but my ancestors calculated that to produce food
in this way would also be expensive. If you were to cost it
properly you would have to count in the cost of the
accommodation used by the animals along with their food.
We could have gathered food for them and brought it in
from outdoors. Once our colonies were established the
robots would have done that for us, but even robots tie up
materials and use energy, so it would not have been free.
In any case, we live indoors, too, most of the time, and we
need all the space we can get just for ourselves. There is
no housing shortage, but we would prefer not to share our
homes with cattle, sheep and goats. Anyway, it would be
unhygienic.

The plants we grow in the open serve to develop the

planet. The plants we need for food are grown indoors, hydroponically. Some are eaten direct, as leaves, fruit, stems or roots. Others are processed to extract their protein, which is flavoured, coloured, and textured, not to make it resemble meat—not deliberately anyway—but to produce food items quite unlike any that exist on Earth. The ingenuity of terran cooks has been inherited by martian food technologists, and developed almost to the status of a fine art. The result is that martian food is different even from terran vegetarian food, and so may require a little getting used to, but once the taste for it is acquired it is delicious and offers great variety.

It is also very nutritious. The nutritive value of food that has been produced in a factory can be known very precisely, and matched to the requirement of the person eating it, and that is more than can be said for food grown in the open. There is no obesity on Mars, nor any of the other ills that on Earth can be traced to bad dietary habits. A few of my fellow passengers will find that despite our low gravity that makes them seem lighter, they will also become slimmer after they have lived for a time on Mars, and they will be healthier for it.

Ruminating thus, I have been interrupted by a call over the public address system asking people to watch the screens. There have been various public information sessions throughout the journey. They are included, I think, mainly as a source of free entertainment. At least, I can think of no other reason for advising space travellers about avoiding accidents in the home. They have mentioned food, though.

Today the 'naturalization programme' has started and as a kind of unofficial guide to Mars I have acquired a mechanical colleague. Being Martian, I do not attend, but then I watched all the arguments on Mars when the programme was being designed and in the course of that I became rather familiar with the content. Terran passengers are expected to watch the screens at certain hours, to receive instruction in what to expect when they land, where to go and what to do. Their attention is held by the simple device of using their own names frequently in the commentary to the pictures, a trick achieved by feeding the names into the commentary program which drives the voice synthesizer.

The aim is simple and limited. This is not a general description of life on the planet, but merely an attempt to avoid confusion. We think it represents progress from the days when immigrants on Earth were herded into large, impersonal sheds, passed from uniformed official to official, told nothing of what was happening to them, were made frightened, confused, and were often humiliated. This programme explains the docking procedure when we reach martian orbit and meet the ferry. It divides the passengers into groups for disembarkation. It describes the reception complex, the processing, and the arrangements that have been made to convey everyone to a final destination on Mars, together with such important details as the workings of the planetary transportation system, the monetary system, ways of buying and selling property, and the community services that are provided.

It is all routine, technical, fairly detailed, and the fact that there was controversy about it makes me think that we Martians may be insecure. It all looks very innocuous now, but some years ago there was hardly anyone on Mars who did not feel strongly about it. We all wanted to be seen at our best, as warm, convivial, welcoming people. Some of us thought the programme should concentrate on our culture— most of it imported, of course—or on the improvements we believe we have made to some of the old, terran ways.

Then we realized that it was a little late for us to be selling the planet to people who were only a few days from arriving. It would be more practical from our point of view, and more useful to them, simply to explain what was about to happen. If we could ensure that everyone knew where to be and when, and where to go next, we would have achieved quite a lot.

In fact, the programme tells people a good deal. For the first time they believe they are being addressed directly by Martians who appear to be in authority, who speak to them as fellow citizens. They acquire status in a martian context and this is very important to them. My friends tell me it makes them feel different. They find, they 'belong'. It was not planned in this way, but by using personal names, by demonstrating to each individual that he or she is expected and a place has been prepared, this first real impression of Mars is a very hospitable one.

I had been expecting to meet a group of Terrans to continue our discussions, but that will have to wait. I will let my binary friend get on with it. While others are watching their screens, I have time to relax and read. I read on a screen, of course. The luxury that impressed me most on my visit to Earth was the abundance of books. There were books everywhere. Every town, every village, had a library, or so I was told, and when I was invited into private homes, I found many Terrans have books of their own.

On Mars, we have every book that was ever written, freely available to everyone, and most of the journals and magazines as well. There is no work of literature, no scientific paper, no historical document, you cannot obtain whenever you want it. But books, as objects, are extremely rare and very few of them are owned privately. Our terran literature was imported, as it were, by radio. It came as microwaves packed with data, to be stored in electronic archives, all filed neatly to make it easily accessible. We were able to tell our terran friends of the existence of documents they did not even know they possessed, because the actual transmission was organized by technicians, rather than by academics. We just contacted a library, such as the Library of Congress, or the British Museum, and asked for everything. Technicians then worked through the stock, item by item, placing books and manuscripts in a machine that turned the pages, scanning each of them as it did so, and converting the picture into radio signal. The technicians did not need to understand the literature, only lift it to the machine and then replace it on the shelves. Then, when each library told us we had received everything, we asked for the basement to be searched. That is where some of the treasures were found, amid rather a lot of dross.

We have everything, we can see anything we wish whenever we wish, and we see it as it appears in the original because we are receiving pictures rather than words, but we cannot handle the books themselves.

I think that in a few years this situation will be remedied, at least in respect of martian literature. At first we lacked the raw materials for the making of paper. We could have used the plants we grew as a source of fibre, but they had a much more important job out in the landscape, and the land needed them just as much when they were dead as when they had been alive. We could have synthesized a substitute for paper, but it seemed hardly worth it when we had access to the information in a convenient form. There were more important things to make. Now, though, the desire to possess books, to handle them as objects, is making itself felt, and a few are being manufactured.

I do not wish to give the impression that our lack of books implies any lack of creativity. Not everyone shares my love of books as physical objects, even on Earth. Martians use simulators, just as Terrans do, with which we can experience adventures, feel emotions, that provide a kind of literary experience but one that is available to far more people. What once required a combination of an imaginative author with a great mastery of language and an imaginative, sensitive reader, can now be achieved by a machine. I can read what a terran author may imagine it was like to live in former times, for example, or I can see and feel it for myself. I can hear the sounds, smell the smells, feel the wind on my face and see all around me the world into which I have been plunged. What is more, I can act in that world. I can behave in any way I choose, and those apparently real people and objects I see will react to me. I can feel the kicks and the caresses! Simulators have encouraged everyone to be creative. But they are not books.

For much the same reasons, we have never used money on Mars, if by money you mean metal tokens and paper bills. We have metals, of course, in abundance, and I dare say we could have devised a substitue for paper money—and one that would have lasted longer than the old bills they once used on Earth. It was just that the materials could be put to much better use, to provide things people really needed. As a physical entity money has more or less fallen from use on Earth now, but it is not so long since terran immigrants were a little frightened to find themselves without jingling purses and bulging wallets.

You do not need money once you have a complete network of electronic communication and machines that can store numbers and manipulate them. To make a purchase you use a video screen, asking it to show you a selection of items from which to choose. You make your choice, and the system that lies hidden behind the screen does the rest. It has identified you, so it needs only to inform the robots in the appropriate stores, who arrange delivery and reordering, and the central bank, which deducts a number from your account and adds it to the account of the supplier.

Regularly recurring bills can be paid automatically, without involving you at all, but there are not many such bills. Terrans are still surprised at the services that apparently are free on Mars. No one pays for accommodation, for food or clothes up to a fixed value per week or year respectively, for education, for medical treatment, or for a whole range of other necessities. This is not because martian society is especially benevolent, or rich, but because it is pointless to carry out millions of small numerical transactions that are so repetitive. Such transactions used to lead to endless problems on Earth, where people used to be given money only in order to have it taken from them at once. Since the transactions concern necessities, everyone must have them. However, people who were given insufficient money could not acquire them in the usual way and so were 'in want'. This meant their needs had to be met in a different way, either by supplying the goods directly or by giving them the money they needed. It was clumsy, inefficient, and often unjust. It is much neater, and more convenient, to give people what they must have and allow them to use their money to buy goods beyond the defined minimum. Perhaps this means we are paid lower wages than people might be on Earth, but the difference is only apparent, not real.

The general principle of avoiding unnecessary transactions applies throughout Mars, but details vary widely. The Daedalians, for example, claim descent from the most fundamentalist of the 'green is beautiful' sects and charge for all food that is not green in colour, a policy that has produced curious fruits. In Daedalia your hardly ever see green food. People colour it as they prepare it, and pride themselves on the most original cuisine in the world. It is certainly the prettiest!

We do have wages, of course, and although we do not use money we do have currency. Our unit of currency is the talent, a name used by the Assyrians, Greeks, Romans, and other ancient Terrans, but with an alternative meaning we feel is rather appropriate to the way we value individuals. 'Talent costs talents,' we say when we are demanding a rise. Talents, of the monetary kind, exist only as numbers in data banks, added and subtracted by computers, and although there are few local variants, the unit is acceptable to computers everywhere.

When we are able to buy genuine martian books, talents will be deducted no doubt. Mars has had its own authors from the very beginning, as well as its own musicians, painters, sculptors, actors, indeed artists and craftsmen of all kinds. We do not consider the arts a luxury, any more than we do the sciences. They are essential for the healthy functioning of any society and will emerge by themselves. That being so, those who planned the first martian communities felt it advisable to establish a high standard from the start, and they also provided good training for would-be practitioners. There are colleges of music, drama, and fine arts on Mars and such is the stimulus artists find on our planet that this aspect of our culture is developing rapidly. I have heard that my friend the drama director is to be sent first to the drama academy in Memnonia, from where he will be able to travel to Olympus without too much difficulty. It is not so very far.

So I read, but my task here is not finished. There is more, much more, to tell.

You may remember that through the authority invested in him by the United Nations, Sir Travers Foxe effectively controlled Mars. He was in a position to bargain.

When it became known that the climate of the planet had been changed, that the place was potentially habitable, the Foxe headquarters were flooded with applications from prospective migrants. At this stage, however, there was no means of conveying them to Mars. The rockets that had been used to carry CFCs had been used and in any case they would have lacked the power to carry human passengers. New and larger vehicles were needed for an operation that had to be planned with great care, to the last detail.

At that time the United States and the Soviet Union had large stocks of long-range, liquid-fuelled rockets that had been designed to carry warheads. Like the solid-fuelled rockets, many of these were becoming obsolete. They did not present the disposal problem presented by the solid-fuelled rockets, because their fuel was kept separately from the 'hardware', but there were many of them and the two governments were not reluctant, in principle, to part with them for the Mars enterprise. Negotiations began.

Foxe was a clever negotiator, expert at releasing just enough information, as leaks that could not be tracked back to him, to play one side against the other, so that before long the two nations found themselves manoeuvred into competing for his custom. He did not neglect, of course, to point out to the officials with whom he dealt the overwhelming service he was able to perform for their masters.

The world wanted disarmament, desperately. The rea-

sons for the arms race were complex, but it had acquired a
terrible inertia that made it difficult to stop. In the west,
entire industries depended on its continuance, and those
industries represented capital investment, employment, and
votes. In the east, where both capital and labour were in
short supply, the incentive for stopping was greater, but
was overbalancd by the fear of attack. It seemed to Foxe
that if the west could be induced to disarm, the east would
be bound to follow, and he was able to overcome the
largest single objection. He could provide an alternative
outlet for the products of the armament industries, at least
until they could devise alternative products and markets.
As it turned out, they did not need radically new products.

He persuaded western governments to supply him with
rockets and their fuel, and to do so with a great fanfare of
publicity. The arrangement was presented to the world as
the conversion of swords into ploughshares. It was what
the world wanted to hear, and the move could not have
been more popular. Never had a United States administration won such immediate, unqualified, world-wide support. The Soviet Union responded at once by providing
more rockets and technical assistance in modifying them,
and Soviet citizens were permitted to apply to migrate.

It was about then that the manufacturers realized that an
entirely new market was opening for them. Already they
were involved in space enterprises. The Shuttle orbiters
were being flown by private corporations and governments
no longer had any monopoly over the launching of satellites, but a few calculations showed that the opening up of
Mars to human colonists would require an endless supply
of the equipment they were making. Ships would be
needed, and their engines, guidance and control systems,
ground control stations—the full list was very long indeed.
It meant that the armaments industries could abandon the

unpopular manufacture of weaponry and commence the popular manufacture of 'tools for tomorrow', which just happened to be more or less the same things.

Everyone was happy. The situation was unprecedented in modern history, and perhaps in all human history. The tensions that had led to the arms race were relaxed and relations between east and west entered a new phase, first of tolerance but then of actual friendship. Peace movements closed down, their objectives having been achieved. Ordinary citizens were happier, and seemed to like and trust their politicians. It was not long before the obsolete rockets were followed by ones that were not obsolete in the sense of being outdated, but that were redundant because the world really was disarming.

There was only one doubt. How was the Mars venture to be financed?

The transformation of the martian climate had given the planet a monetary value. It had become real estate, land that could be bought and sold. Those who wished to migrate were asked to buy land on the planet. At first they could not stipulate the precise plot they would own, but they could be supplied with documents entitling them to a site whose area was determined by the money they had paid. Land values on the planet were fixed at a minimum, but then responded to supply and demand. The supply was limited, because there was no plan then to colonize the entire planet. Indeed, the whole planet has not been colonized yet. The most favourable region for habitation lay in a belt in the northern martian tropic, and excluded all the higher ground, so that mountain ranges were omitted. This gave land something of a scarcity value and supported the price. The money paid was used to equip the expeditions, the whole financial operation being supervised by United Nations agents. In effect, for the price of the land that one day they would own and on which they would live, people were buying a share in a ship, its fuel, and the supplies that would get them started in their new lives.

That was how the custom began of supplying necessities, including housing, free of charge. People had paid for their accommodation when they bought their tickets. They could not be charged twice. Similarly, they had paid for the seed and other materials they needed to produce food—on land they had also bought. All necessities, for the rest of their lives, were paid for before their departure. It is rather different now, but the custom has persisted because, as I said, it is convenient.

From the very first, therefore, Mars was largely self-financing. Because they could take so little with them, people had to sell everything they owned on Earth, so they went to the immigration authority with the value of their homes, farms, factories, stocks and shares, in their hands, and with their own labour and skills provided free. It was more difficult for people from the 'command economy' countries, where property was not owned privately, but their governments awarded cash sums to certain of their citizens that were equal to the sums their fellow migrants could raise.

The enterprise cost far less than anyone imagined. For years it had been supposed that space ventures were necessarily expensive, and in the early years so they were. They were expensive for two principal reasons. The facilities that were needed, on the ground as well as in space, had to be invented, tested, and built, all from scratch, before the space operation proper could begin. The high cost included the whole of the research and development that backed it. Because everything was so new, and because everything was paid for by governments, safety was a paramount consideration. Governments, it was said, could afford to spend money. What they could not afford were catastrophes involving the loss of human life. There was only one way such catastrophes could be avoided when all the equipment was new, and that was to duplicate, triplicate, quadruplicate everything. This added cost, and it added launch weight, which added still more cost.

As time passed, more and more experience was gained, the necessary installations were tested in real operation very thoroughly, and costs began to fall rapidly. The research and development had been paid for by governments, and it did not have to be paid for again.

The big change came, though, when private organizations became involved. A private company, like a private individual, cannot afford to spend very large sums of money, but it will not be destroyed by a single catastrophe. It is not that the companies economized on safety to the extent of taking risks, but they achieved safety as cheaply as they could, mainly by using only equipment that had been tested and not permitting it to exceed its design performance. The difference was rather like that between a commercial airline and an air force. The airline flies standard aircraft, on pre-arranged routes, and no aircraft is taken anywhere near its performance limits. Military air craft are meant to be flown at their limits, and to have those limits extended all the time. An airliner carries many passengers, a military aircraft very few, and so the cost of the machine itself can be related to its human payload. As a result, military aviation is much more expensive and also more dangerous.

There had been strong hints that the cost of the Mars project would be far lower than had been estimated at around the time the first humans were visiting the Moon. Then numbers had simply been extrapolated, and a visit to Mars had been treated as 'so many times' more expensive than the visit to the Moon. As I just said, this was unrealistic, because almost everything needed for the Moon project had to be designed and tested specially. In 1979, Freeman Dyson (in *Disturbing the Universe**) reworked the sums, not in terms of Mars to be sure, but in terms of a kind of island colony that might be built in space. This really would have been expensive, especially compared with the colonization of a planet that existed already. Nevertheless, his approach was interesting because he related the costs of space colonization to the costs of the *Mayflower* migration to North America, and to the Mormon crossing of the United States in 1847, by translating records of the accounts kept of those enterprises into 1975 monetary values. He found that the *Mayflower* expedition cost each family a sum equal to the wages a person might earn by 7.5 years of work. The Mormon expedition, on the same basis, cost 2.5 years of work per family.

His 'colony' would have cost 1500 years of work per family, although that never was realistic, but another project, to 'homestead' asteroids, might have cost only six years. In 1982, James Oberg (in *Mission to Mars**) calculated that a manned expedition to Mars, with five astronauts, might cost about $20 billion, which is a little more than it cost to develop the Shuttle orbiter, and less than one-third of the cost of the Moon landings. Once it became clear that the equipment could be produced in quantity, that to begin with the launch vehicles could be supplied and fuelled free, and that the travellers did not plan to return to Earth, the figure of $20 billion had to be reduced substantially. Costed over all the equipment for all the journeys, and divided by the number of families travelling, the final price for a ticket to Mars dropped to the equivalent of three of Dyson's 'man-years'. In other words, at average wages prevailing at the time, it would have taken

*Harper & Row, New York, 1979. Pan Books, London, 1981.
*Stackpole Books, Harrisburg, Pa, 1982.

one person three years to earn the price of a ticket for a family of two adults and two children. This is a way of calculating cost that makes it seem homely, but of course no one could work for three years and spend no money at all on living during that time. Nor did anyone need to do so, because the value of the family property covered it more than adequately. Mars was more easily attainable than anyone would have dared to suggest a few years earlier. It just goes to show that when people are determined to do something, they will find a way to do it.

After the first few years, costs fell yet again. This time the reason was purely commercial. The one spaceline that had been licensed to carry passengers to Mars was suspected by some people of making excessive profits. It lacked competition, and fixed its own prices. When its monopoly was challenged it was in a weak position to retaliate. People were offered fares one-third cheaper if they would accept a lower standard of catering and accommodation during the journey. The new, strictly speaking illegal, spaceline attracted customers. In fact it expanded the space travel market to a whole new range of customers who until then had been narrowly excluded by price. The new line was popular, and soon it received its licence.

The new spaceline thrived for a time, but the relaxation of standards it introduced was soon exploited by other lines, offering still lower fares for still less comfortable travel, and some of them began to economize on every aspect of the operation. That is how the appalling migrant ships came to operate, and they went on operating for some time before the survivors among their former customers, who were now Martians, managed to stop them. Still, each time the price fell, more people were able to travel.

*   *   *

Money was not really important. How can you put a price on an adventure of this kind? It seemed to some people at the time to be much more important to ensure that only the 'right sort' of people were accepted. That proved immensely difficult.

If you are recruiting astronauts, it is not too difficult to draw up a job description to guide you. If you are recruiting migrants it is not so easy. You need some specialists, of couse, some 'experts', but you also need people whose skills are not immediately, or perhaps I should say obviously, relevant. Anyone who is fit can be useful. It takes little skill to move rock for building or smelting, to fasten bolts or rivets, or to work switches, and when you try to draw up a list of the different skills a new colony needs, the list soon grows very long. Basically, you need strong arms and backs, and quick wits. Once Mars was colonized all manual work would be done by robots, but this was not so in the early days. The robots were available, to be sure, but they could not buy tickets for themselves. Spaceship passenger payloads had to be maximized. The migrants could have built robots of their own when they arrived, and later that is what they did, but at first there was no time. Shelters had to be built, plants sown, essential materials supplied, and only humans were around to do the work.

Obviously you will need people who can work as builders and stonemasons. You will need engineers of several kinds. You will need plumbers, electricians, computer programmers, geologists, chemists, and farmers, but they still need help. You will need cooks, doctors, statisticians, cleaners, and clerks. People must be kept informed about what is happening, and their natural curiosity and wish to learn about their new companions mean that journalists are needed, and the technicians that support them by constructing and maintaining the communications systems. People married, and within a year babies were being born. Priests were required. Other children arrived from Earth with their parents. We do not have schools and universities in the old terran sense, but young people had to be educated, right from the day the first of them arrived, and provision for education had to be made and the educational programme planned. So it continued, with the net of our requirements being cast wider and wider. We had to have artists, actors, and authors as I mentioned earlier.

After all, if the colonization of Mars were to succeed, complete communities had to be established. They had to be constructed, as it were, from imported human 'ingredients'. There was no time to replicate terran history and evolve them! Mars needed people of every kind.

I realize this makes it sound as though people were chosen on some kind of professional basis. They were not, and although there were no robots, not all the workers were human. There was ample computing power because that did pay its fare by being used on the ships. The computers could and did provide as much expert advice as was needed, and some tasks they performed unaided. They were the teachers, for example, that provided education for the young. The most important human qualifications were enthusiasm, a willingness to work, and a willingness to work at anything, to be a master of many trades.

Mars needed people of every kind. Or did it? How were the misfits, the criminals, the political or religious fanatics to be excluded? Who were the right sort of people? It is a matter that caused much scratching of heads, for no matter how you describe the kind of person you would most like to have for a neighbour, you are almost certain to end up describing either your own idealized image of yourself or some other, equally non-existent saint.

People were examined carefully, or fairly carefully. They had to be physically and psychologically fit, of course, to withstand the rigours of the journey and of life on the planet. A good standard of education was not essential but it helped. No one was accepted with a criminal record. Beyond that, the migrants selected themselves rather well, first on Earth and then on Mars.

Would I gamble all I possess, my life, and the lives of my family? Am I the stuff of which space pioneers are made? I have no opportunity to do so. I have never had to face the challenge, and I do not know. What I see very clearly is that this requirement, which was imposed by the very nature of the project, defined a certain type of person and excluded those attracted by immediate economic gain for a minimum of effort or commitment. That was the first stage in the selection process, and it forced people to select themselves.

When people arrived on Mars a second self-selection took place. Some accepted the situation more readily than others, but in the end everyone accepted it. On Earth, there had been those who imagined that the martian colonies would be rather like the terran colonies of old, that within them all the old rivalries, prejudices, envies and hatreds would continue. They imagined conflicts emerging based on ethnic or class origins, on religion, political belief, or customs. They feared that a new rich would oppress a new poor, only to be overthrown be revolution in a lawless, ungovernable state. Should national rivalries appear again on Earth, which was always possible no matter how euphoric international relations might seem, these would be translated to the new planet, where armies would be recruited, and wars fought on martian soil in the name of terran tribalism.

It was not like that. The analogy was quite wrong. Mars was not like America, or Australia. It was like nowhere on Earth, but if a comparison were needed, perhaps Antarctica might serve, inadequately. Antarctica has no permanent human population, but people do live there. Those people come from all parts of the world, from countries that in the old days were on the verge of war with one another or even, occasionally, they have come from countries that were actually fighting. While they are in Antarctica, however, they co-operate. National rivalries are forgotten. They help one another. They have no real choice, because without help a human being in difficulties is very likely to die. Altruism is the most sensible behaviour and its benefits are clearly apparent. It is a case of 'do as you would be done by—or die'.

Mars was like that, not because the climate was so harsh, but because the air was unbreathable. While they remained in their buildings, or had their breathing apparatus with them, people were safe, but should anything fail they would be in immediate and grave danger. From the very first, each person learned to give help automatically wherever it was needed, in the certainty that such help would be reciprocated. The robust individualists were poorly equipped for survival, and either they realized and accepted the fact of their own accord or, sooner or later, they needed help, received it, and learned interdependence that way. Death was always, and easily, available as an alternative.

For much the same reason, wars were not fought on Mars. It was naive to suppose that terran governments could imagine their writ running over so great a distance.

The cost of transporting armies and their weapons might have been tolerable, but without the support of Martians life on the planet would have been very difficult for them. People changed their loyalties when they changed planets and there was no guarantee that soldiers would be given hospitality or even that they could seize and hold the life-support they needed. Sabotage would have been too easy where food was grown inside buildings that could be poisoned, and where the very air could be stolen.

Loyalties changed, old tribalisms were lost fairly quickly in a new environment, by people who had grown up on Earth in a fairly internationally minded age. People were hospitable, as desert peoples are hospitable on Earth. Yet this alone would not have preserved the peace. Tribalism soon asserted itself in a new form.

It was not long before colonies began to differ one from another. The people who comprised them seemed almost to go out of their way to invent differences. Perhaps there is an evolutionary value in such diversity, so that disaster may be dealt with in different ways by different groups of people, some successful, others not. The differences expressed themselves in customs, in styles of dress and ornament, and in local variations of speech. In some cases old terran languages were preserved, to provide a kind of cultural base for divergence, but quite new differences developed and became more important as the years passed. Inevitably, differences led to rivalries, and there were disputes, but the disputes were resolved in a curious way.

On Mars there was always room. No one was crowded, and resources were plentiful once the first few colonies were well established. When diversity led to confrontation, small groups of individuals would break away to begin a new settlement. A strongly held opinion after all, can seldom count on total support even within a particular community, and there were always dissidents who could, and did, depart to start life on their own. Sometimes it was the would-be aggressors who left, only to find themselves too few in number to fight, or those of a more pacific persuasion, whose departure weakened the community they left. Tribalism tended to encourage diversification, and the diversification made war more difficult. The existence of tribes led to the formation of more tribes, and so it continued. Today there are so many cultures, so many of these 'tribes', each jealous of an independence that no one else really challenges, that war has become a political and logistical impossibility.

Today, if you wish to fight in order to win territory, then Mars has plenty of territory to which you are welcome without going to the trouble of fighting for it. If you wish to fight in order to impose your will on humans, which seems to be the only reason that makes (limited) sense, then the very diversification of martian society would make it difficult to overthrow.

So, although there were times when it looked probable, in fact there has never been war on Mars. It is the most peaceful of places, which is odd when you remember its ancient mythical associations.

Colonies, you recall, developed in a kind of network. New arrivals would be sheltered and equipped at an established site, and then would move out to the land they would occupy, and would build there. In time, the map looked a bit like a net, with a settlement at each of the points where lines crossed, and further expansion involved filling in gaps in the net and then in making it larger. Each colony was fully self-sufficient and independent. There was planet-wide agreement on matters that concerned everyone, such as the unit of currency to use and its value, and on the type of communications systems, but such agreements were minimal. Currency rates had to be fixed because if money were worth more in one place than in another there might be internal migration of people to settlements that could not support them. Communications systems had to be compatible with one another, so that individual colonies could contact one another, but they must not be allowed to interfere with one another, as they might if several were to broadcast simultaneously on the same wavelength, for example. However, the high degree of autonomy enjoyed by each colony meant there was no central power. Power did not reside in any place, any person, or any group of people. It was dispersed. The Martian Council, consisting of representatives from the colonies, discusses matters of mutual concern, mainly to do with immigration—and my report—but it does not hold power in the old terran sense. It is not a 'government'. We need no such institution, and our administration is conducted by computer program. Thus the 'martian people'—a fictitious entity in any political sense—could be overthrown only if each colony separately were overthrown. That, surely, would be a tedious business.

We do have some crime, naturally, but we have solved one of the difficulties that existed on Earth. With no animals to hunt, no non-human to threaten a human, there was no need to allow guns to be taken to Mars, and their banning was very simple. Weight is such an important factor in a launch from Earth that every item of cargo is examined and has to be justified. The screening has been extremely thorough from the start, and no gun could ever have passed it. That would not prevent someone from manufacturing a gun on Mars, of course, although so far as I know it has never happened, but our final sanction against wrongdoers is final, and awful. Ultimately, antisocial behaviour can be dealt with by the expulsion of an offender from society and even today you cannot live for long in the open. It is a very final kind of rejection. We do not talk about such things. I think the sanction has been used only rarely, perhaps not at all. Most offenders are dealt with by means of economic penalties.

They are coming back from the screens and I did not manage even to switch on my own screen and invoke a book. My reflections have prepared me for the keeping of promises. I can explain, I think, a little more of the way the martian colonies arose, and why we collaborate among ourselves rather than fighting.

I have said nothing about transport, though, and I must, for that will delight Terrans. So far as I know it is not even mentioned in any of the simulator programs. No one seems to think it important to tell of the way we travel about our planet. Martians take it all for granted. That will be a treat to come for people who are unfamiliar with modern airships.

Airships
and swamps

Last night we arrived. Preparations began yesterday afternoon. The crew were busy everywhere stowing equipment, and meals were somewhat rudimentary. Passengers were advised to pack their belongings except for items they needed overnight. It was good advice. Yesterday, things lay in neat piles where they had been put, while their owners packed them one at a time. Today nothing stays where it is put. Toothbrushes, night clothes, float free, and I saw a middle-aged and formerly staid gentleman drifting down a corridor in pursuit of his trousers.

We slept strapped to our beds, and while we slept the ship ceased to rotate as it entered its orbit of Mars and prepared to dock with the first of the ferries. For the moment we are weightless, and so is everything else. The chaos around me demonstrates rather eloquently the ease with which people forget what they learned so recently about coping with weightlessness.

I have a special status and that means I must travel on the first ferry. I have said my farewells, sadly, because I have made several friends during our voyage and I know, even if they do not, that it is most unlikely that we will meet again. I know what awaits me. I will be met, hurried to the transport that will be waiting, and taken to the people to whom I must give my report. Then, my mission completed, I will go home to Zephyria, to my family, to my old life, and to my memories. I do not expect I will leave home again, at least for long journeys, and the newcomers are bound for the more recent settlements, not to mine. When they arrive they will be fully occupied in adjusting to their new lives and by the time they feel established I dare say they will have forgotten me. They could call me, of course, so we could see one another if only on screens, and could talk, but they will not. So, farewell.

Yet, once home, once the excitement of my return has waned, perhaps I will read through these notes. Perhaps I will edit and expand them into a document that may be of use to others who must make 'the big jump'. Who knows but that such a text might persuade the terran authorities to include more information about Mars in their educational software. So, for the time being, I will continue. I will bring my story to its end. Perhaps, because of that end, it will become a story of some historic interest not so much for its content as for the circumstances under which it was written.

Everyone will be taken first to the Aram spaceport, not too far from the monument marking the place of Chryse where one of the first unmanned terran vehicles landed. Terran passengers will stay there long enough to have a meal, then they will depart for their final destinations, and their first real views of the martian countryside.

They will travel by airship. On Earth, my terran friends were surprised to learn of the extent to which we rely on airships. They are used on Earth, of course, but there they are only one form of transport among many. On Mars they provide the only means for making long journeys. We have wheeled vehicles, but they are used only within settlements and there are no roads linking one settlement with another.

The reason for our love of airships is partly historical. When the first settlers began to arrive they needed to move about the planet and to carry cargo, in order to establish new settlements. There were no roads, of course, and so wheeled vehicles were of limited use. They existed, but they were slow and they tended to break down, especially in very rugged terrain. Their range was very limited, too, because they were powered by batteries that needed regular recharging. Some form of air transport was needed.

You could build fixed-wing aircraft on Mars, though probably not helicopters. The air is thin, so the wings would need to be very long and narrow. If the wings are fixed this is possible, but if they must rotate as do the rotors of a helicopter, the longer they are the greater are the engineering difficulties that arise with them. Fixed-wing aircraft would fly, but just as there were no roads, so there were no runways either. My forefathers might have built a runway for themselves, and some aircraft to use it, but where could those aircraft land, except at the single runway? How could they move people and supplies to new localities if they could not land and, in the martian air, they could not even parachute their payloads to the ground? Aircraft were even less satisfactory than wheeled vehicles.

Airships provided the answer. They need no runways,

no airfields, nothing in fact, apart from a small, level area on which to set themselves down. From the first we have used hydrogen as a lifting gas. It is plentiful, cheap, and on Mars hydrogen is even safer than it is on Earth, because in martian air it will not burn. There is insufficient oxygen in our air for there to be fires of any kind.

They were easy to build, simple to operate, and they could be made large enough to carry very heavy loads. Also, they can fly high enough to take them across our more mountainous regions. The ceiling for a lighter-than-air craft is determined by the density of the lifting gas compared to that of the surrounding air, and on Mars air pressure decreases with height much more gradually than it does on Earth and, as I mentioned earlier, the atmosphere extends much higher than does the terran atmosphere.

The airships look not so much large, as fat. They are broad in order to offer a large surface to the wind, for they are powered mainly by the wind. On the routes they usually follow this is convenient. The wind pattern on Mars is much like that on Earth. As oxygen began to enter the air, an ozone layer formed, and the atmosphere became divided into a distinct lower layer, the troposphere, and a stratosphere above it. Within the troposphere, warm air rises at the equator, spreads toward higher latitudes at high altitude, then descends as cooler, drier air over the subtropics. This gives Mars its most humid climates close to the equator, with very dry deserts to the north and south, and a further, somewhat moister, climate in the mid latitudes. It also produces trade winds to either side of the equator, and since our air routes are mainly east and west, the airships use the winds.

They have auxiliary power as well. They need this to help with docking, to make it easier to change direction —you cannot manoeuvre sails—and to drive them should they enter the martian doldrums. The power for this is supplied by the sun. The big, fat ships have arrays of solar cells on their upper surfaces. These produce enough electrical power to keep batteries charged, and the batteries drive the electric motors, which turn large propellers.

Recently, some engineers have been experimenting with a design they think might be an improvement on the traditional airship. This is a ship with a sail, but a sail that is cylindrical in shape and carried vertically below the main part of the ship. The cylinder rotates on its vertical axis, driven by the solar powered motors, and the rotation causes a difference in pressure on the surface of the cylinder, which drives the ship. Perhaps one day we will ride in airships driven by cylindrical sails exploiting the Magnus effect.

Martian skies, then, are crossed by large airships, but even with so delightful a vehicle, journeys are not undertaken often. Most of the ships you see are carrying freight. People sometimes travel just for amusement, especially when they are young, but there is no real need to travel anywhere far from home. We maintain contact with other settlements almost entirely by electronic means and an increase in air traffic usually means a new batch of immigrants has arrived.

From Aram, I will be taken in a small airship north-east to one of our more northerly settlements, in the McLaughlin Crater, in the Indus region.

Most people will start their journeys by travelling more or less due east or west, turning north or south later, when they are close to their destinations, and they will set out by day. If they arrive late in the day they will stay at the spaceport overnight and depart the next morning, so that when they really see Mars for the first time, they will actually see it. Lacking a large moon, martian nights are dark. They may not complete their travels in a single day, of course, in which case they must continue into and perhaps through the night, but at least they will have had one, breath-taking view, with a martian sunset and dawn thrown in as a bonus.

Those who go east will pass Edom, and the crater named after the astronomer Schiaparelli, who discovered the 'canals' of Mars, and into the crater fields of Aeria, rising fully six kilometres above the surrounding terrain. From there, if they continue, they will descend gradually into Syrtis Major with, if they are lucky, a distant view of the peaks of the mountains of Hellas to the south. Beyond that again, eventually they will reach the smooth plains of Zephyria, where perhaps children whose names I know will wave to them.

Those who go west, across Chryse, will enjoy more spectacular scenery. Over Chryse, their ship will turn south so that they may pass along the biggest canyon system on Mars, separating the regions of Tithonius Lacus and Ophir in the north from Syria and Sinai in the south. About four thousand kilometres from end to end, in places seven hundred kilometres wide, and up to six kilometres deep, with a maze of side branches most of which are much bigger than the Grand Canyon on Earth, the Valles Marineris is one of the wonders of the solar system.

The Valles is inhabited, which adds to the wonder, for it forms one of the most fertile regions of Mars. Not far from its western end the land rises steeply into the Tharsis Montes, dominated by the three volcanic mountains Arsia Silva, Pavonis Lacus, and Ascraeus Lacus, forming a straight line with the central one opposite the branching canyons at the mouth of the Valles. Behind them, and rising to a height of twenty-three kilometres, is Olympus

Mons, known to early terran astronomers as Nix Olympus, 'the snows of Olympus', because they could see it from Earth.

The travellers will leave the Valles heading north-west to pass by the southen side of Olympus and they will have an excellent view of it. More to the point, so far as Martians are concerned, the entire system drains into and through the Valles. It is a place where the Coprates River flows wide and deep, into the Mare Erythraeum.

Like other mares, the Mare Erythraeum is not an ocean, or even a sea, in the terran sense. It is a place of marshes, swamps, and large but shallow lakes that on a fine day shine pink beneath martian skies. It is where reeds wave in the wind and sedges form clumps like little islands to serve as stepping stones. It is where children play, and where families bathe on warm summer days and sail their small yachts.

There are rivers on modern Mars, a few marshy areas, some lakes, but to experience our world, to feel the planet as it were, you must stand alone, quite still, and absorb its atmosphere. The green hills rise away, with patches of bright lichen colour where large rocks crop through the thin soil, to the red martian soil on the higher, more exposed ground. Here and there trees murmur quietly in the thin martian wind. Perhaps a stream splashes over loose stones, rattling them, or a small waterfall utters a steady but muted roar, but the sounds are of inanimate things. No bird sings, no insect buzzes, no cow or sheep cries it presence. And above the scene, white clouds move at a leisurely pace across the palest of pink skies. The landscape is at peace. It rests. It awaits its future as it waited billions of years for the arrival of this much life. It is patient, as it always has been patient.

When I stand alone amid the Zephyrian fields, I sense deep, deep tranquillity. Terrans tell me they have this sense on Earth, but I do not believe them. I tried it. I went out of their houses, to stand on Earth as sometimes I stand on Mars, but peace was not there. There were sounds of birds and animals, some of which might be dangerous for all I knew. There was traffic noise, and aircraft noise, and human voices. Terrans could not hear such things I dare say, for what passes among them as silence to a martian ear is din.

Perhaps, in a terran desert, there is such peace to be found, that deep stillness which brings a tranquillity that penetrates the very core of the human spirit. At least desert peoples have always said so. On Mars, such peace is everywhere, but there is a difference. On Mars, it is the peace of life developing, not of life held in check, frustrated, as it is in a desert.

The settlements are not visually intrusive, they cause little disturbance of the landscape for they have no roads or railways leading from them and no cultivated fields surrounding them. They consist of buildings, made from local materials, and partly buried so that the green vegetation seems to crowd in upon them. True, there may be many such buildings in a large settlement, and there are roads linking the buildings themselves, but yet the settlement is self-contained within the countryside and unless you know in advance precisely where it is, a settlement can be difficult to find. There are no visual signposts, no clues that can be seen to lead the traveller from miles away, only invisible radio beams and, as you come closer, a telltale radio mast. Long-range radio communication uses satellites parked in geostationary orbits, of course. Space is part of our heritage, and we use it as readily as we use the land or the air.

Unlike terran cities, martian settlements become totally invisible by night. We have no street lights, and our buildings are shuttered tightly after dark to prevent the loss of heat, for martian nights are very cold. There is water, as I have said, but compared to Earth, Mars is very dry. There are no great oceans to absorb heat, store it, and release it when the air is cold, and so to moderate temperatures. The whole of Mars has what on Earth would be called an extreme kind of continental climate, with hot days but bitterly cold nights.

There is one curious consequence of our low air pressure I have omitted to mention, for it is something we take for granted. I think of it now only because I heard a Terran mention that she would welcome a cup of tea. We are about to board the ferry, and so she must wait until we reach Mars, but if she should see the tea being made she may be shocked, for we boil the brew for a full ten minutes. We have to do so, because on Mars water boils at 50°C. That is rather hot for a bath, but too cool to dissolve the vital juices from the dried leaves of the tea plant, or from the roasted and ground seeds of the coffee plant come to that. To produce a satisfactory drink, we must boil the brew!

Incidentally, you cannot destroy micro-organisms, and so sterilize things, by boiling at so low a temperature. You can sterilize things under pressure, of course, in an autoclave, but I wonder whether just from force of habit the early immigrants tried to sterilize things by boiling, and so exacerbated the disease epidemics that caused havoc among them?

So we landed and, as I predicted, a small group of people met me and escorted me from the terminal building to an awaiting airship. The Terrans I left in the reception lounge were starting to queue for the issue of their outdoor breathing apparatus, and those that had received them were studying them quietly, as though trying to decide how they should be worn. Everyone was quiet, the Martians because they were busy, the Terrans because, inevitably, they were awed. A meal will help them relax. It is why they are given one. That lady will have had her cup of tea by now.

That was the last I saw of my fellow passengers. As the airship lifted me away the second ferry was landing and the first was preparing to take off again for another human cargo. It was an ending and a beginning.

For me, too, an end was close. I had been ordered to deliver in person the report I carried. My escort glanced at the bag containing it, but it was not mentioned. They had no idea of its content, only of its importance. We talked of everyday things, of the entertainments I had missed while I was away and, casually, of my experience of Earth. They seemed not much interested in Earth. It is, for them, a remote world. They will not visit it, they have little contact with it, their lives are here, where they were born. It is enough.

I am writing this as best I can in the airship that carries me home from McLaughlin. The ship has been chartered by the Council especially for me. A steward plies me with food at what seems like half hour intervals, now and then I relax and watch genuine martian entertainment on the simulator set in the back of the seat in front of me, I doze, but mainly I reflect on my meeting, and my report.

I was complimented, indeed flattered, for the efficiency with which I handled the mission entrusted to me, and I dare say that in due course I will be awarded a conveniently large number of talents that will help to pay for the enlargement of my home, but that was all. The bag attached to my garments was detached in the presence of witnesses, then taken from me. I did not even open it myself, although it was checked in my presence so that I could agree it contained everything that had been placed in it back on Earth. I was taken before a high official of the Council. She thanked me, complimented me as I said, then dismissed me. I was shown to a restaurant, then to a bedroom where I slept, and this morning I was put aboard the chartered ship.

It was all a bit of an anticlimax, but soon I will be home, among my own people, my own friendly surroundings, and then I will be pressed to talk of my adventures. As I talk, so my memories will formalize themselves, taking on something of the quality of legend. With repetition, descriptive patterns, the recollection of emotions, will become fixed, immutable, and to that extent erroneous.

That will not be imporant. What matters is that my first thoughts on returning to Mars be recorded while they are fresh. I have known the content of the report since before I left, but now, with the conclusions of the report confirmed on Earth and that confirmation passed on, for the first time I am alone in my own world, and have time to reflect.

I must write of the report.

Martians

The report states, baldly and boldly, that we Martians now comprise a distinct species within the genus *Homo*. We are different from Terrans, biologically, and in the years to come we may expect the differences to become more pronounced.

In the old days Terrans used to imagine they had problems with what they called racism. We have brought new meaning to that much abused word. It seems to me that we may face very real difficulties.

The human (we remain human, I insist, even if we are no longer of the *sapiens* species) colonization of Mars produced a classic evolutionary situation, circumstances of textbook perfection in which a new species may be expected to emerge.

I have explained the way in which settlements were distributed. Each new batch of immigrants was taken first to a reception area, eventually at the Aram spaceport. From there they moved into unsettled territory, and built their own settlement, so that the martian colony as a whole spread in a kind of radial pattern, ever outward until the physical limits for expansion were reached. Once a settlement was completed and the people within it were living their own lives, it was pretty well self-sufficient. There was not much direct, personal contact among settlements, but inevitably I suppose, a social hierarchy developed, with an élete formed from the families that had lived on the planet longest.

The élete developed its own customs and, more important, its own aesthetic concepts. Its members felt they had, to a large extent, 'built' Mars. They felt proprietorial, but not in any physically exclusive sense, for new colonists were needed all the time. It was that over the generations they came more and more to reject terran concepts of beauty and to develop their own, martian standards. To them, the martian countryside was in every respect more beautiful than any terran countryside. Martian buildings conformed to higher architectural and engineering standards than any buildings on Earth. And the people of Mars were more beautiful, more sexually attractive, than the people of Earth. They smelled different, and still do.

Of course, the first Martians were no different from Terrans. The differences were invented by the divergences in taste, and tastes diverged more and more widely. The process was almost perverse. It was as though Martians had listed the characteristics that were considered beautiful on Earth and reversed them. This is not so extraordinary. Standards of human beauty have been revised on Earth many times. People who were thought beautiful in Renaissance Italy, for example, would have been rather plain in twentieth-century America. In this case, though, the divergence in tastes was reinforced by other cultural considerations, some of which were practical. The longer a family had lived on Mars the better adjusted its members were likely to be, the better equipped to cope with martian conditions, and such a family was likely to produce more valuable marriage partners. For one thing, the fertility of such a family was more likely to have been proven, and the ability to produce many children was an important asset. The planet needed people. For a long time the greatest danger was that the colonial population would dwindle, until those who remained were unable to look after themselves, especially when they grew old.

Children were brought up to respect the new standards and since these came, as it were in a 'package' bound up with many perfectly sensible ideals, the standards were not challenged. Within two or three generations the new arrivals from Earth actually looked ugly to Martians. No one wanted to marry into an immigrant family.

New customs developed. People wanted to commemorate the achievement of the first humans to arrive, to remember always the difficulties under which they had lived and worked, and as the years passed this desire gave rise to an initiation ceremony, a rite of passage, an ordeal through which all young men had to pass when they reached puberty. It was harsh, but simple enough. Each youngster was sent out, alone, to walk across the desert from his own settlement to another, on a journey that took him up to three weeks. He could take with him as much as he thought he could carry, but before long he found he had to learn to find water, to make very economical use of his air supply, to be thrifty. Many young Martians have died while trying to pass this test, but no one was required to die. The candidate could return, but if he did, he was deemed to have failed, he was not permitted to marry. It is easy to be wise after the event. I can see now that my ancestors were selecting young men for their degree of adaptation to martian conditions, and allowing only the most highly adapted to breed.

There was a further selection, but this time an entirely accidental one. It came in the form of two epidemics of disease, caused by a viral infection that attacked the respiratory system to produce a kind of pneumonia. The virus was brought in by immigrants, among whom it had remained dormant throughout the journey, but when it struck its effect was devastating. On Earth, pneumonia can be treated without much difficulty, and the risk of death among young, otherwise healthy people is small. On Mars, however, any attack to the respiratory system is much more serious. Many people died, and those who survived were naturally much less susceptible to the disease. Since then the disease has become endemic throughout the population, but we have adapted to live with it, and it with us. It is inconvenient, but no longer serious, and now it is wholly martian. There is nothing quite like it on Earth, and although immigrants are inoculated against it when they arrive, it continues to claim a few lives every year among terran newcomers.

There was also a geographical separation. This must seem paradoxical, since the older settlements were those closest to the arrival and reception areas, but it was this proximity that produced the isolation. On Earth, an immigrant family, and especially an impoverished one, would remain close to the place of arrival, often in the centre of an old, deteriorating city, then move to more prosperous neighbourhoods when and if the family fortunes improved. The improvement caused the move, and the move caused a gradual integration of immigrants into the established, host community. On Mars it was different. When they arrived, families were neither rich nor poor. The distinction did not exist. They were moved, all of them as a group, by the existing host community, to the areas into which they were to settle, and these areas were at the periphery of the colony. They passed right over the oldest settlements. They did not land at them, visit them, or have any physical contact with them whatever. It was a leapfrogging progression that maintained the isolation of the central settlements and ensured that as time went on, new arrivals lived further and further away from them.

In time, therefore, three breeding populations of humans came into being. The Terrans, living on Earth, bred as they had always bred. The Martians, counting themselves as the social élite in their own world, bred among themselves. The third population comprised the more recent martian arrivals, who also tended to breed mainly among themselves although there was some interbreeding with the still more recent immigrants where they were installed in settlements nearby.

The important point is that at the extremes the 'senior' Martians and the Earth-dwelling Terrans were completely separate from one another. No breeding between these groups was possible because they could meet only as a consequence of space migration. Terrans arriving on Mars had little opportunity to meet 'senior' Martians socially, and Martians very seldom returned to Earth. These senior Martians, according to the mores they established, bred as prolifically as they could, and within the large area they inhabited they became very numerous. It was, is, a matter of pride among them, us, to have many children, many grandchildren. What is more, the Martians found the Terrans unattractive, and for all I know the feeling was reciprocated. To this extent, the human population had been split in two.

When this happens, when a population of a species is divided, to form two populations that do not interbreed, the evolution of each of them proceeds along its own paths and in time they may become distinct from one another. There are two possible routes to such a division. On Earth, crustal movements or changes in climate may lead to a physical isolation, some barrier that cannot be crossed. A sea may widen to separate land areas that were formerly adjacent, a range of mountains may be thrust upward, even a large river may flow where no river flowed before, and the groups living on either side of the barrier may be debarred from further contact. Alternatively, within an established population the behaviour of some members may change. Among certain birds, for example, the spectacular courting display of the male may be varied by a few males, who attract certain females but not others, and so two groups may emerge that do not interbreed because members of one are not attracted to members of the other. A physical change may occur. One line of male deer may produce antlers of a particular shape, which attract certain females, and again two groups emerge that do not interbreed because they find one another unattractive. There are many ways in which a population may become divided into two or more smaller populations from a reproductive point of view.

Between Terrans and Martians, two barriers existed. There were the physical barriers of space, and on Mars of the pattern of settlement, and the behavioural barrier that reinforced the geographical one. Terrans and Martians were very effectively isolated from one another from a reproductive point of view.

At its simplest, a species may be defined as a group of individuals whose members breed among themselves, but who do not breed with members of other groups. They may not breed with other groups because physically they are incapable of doing do, or incapable of doing so and producing fertile offspring, or because their courtship and mating behaviour makes such mating impossible. They may be sexually receptive at different times of year, for example, or attracted to different kinds of courtship behaviour.

In practice, it is a little more complicated. On Earth I dare say marriages between Kalahari Bushmen and North American Indians are fairly uncommon, but the two groups continue to be members of the same species. The aboriginal peoples of Australia remained separated geographically from all other humans for thousands of years, but did not evolve into a new species. Something more is needed, and dramatic though it may seem the isolation of these two groups of humans is not complete because of all the intermarriages that take place among humans all the way from southern Africa to North America. One or both of the populations must be subjected to a local pressure that selects from among them, and allows to breed most prolifically, those individuals who are best equipped to withstand that local pressure—who are best adapted to the circumstances under which they live.

Speciation is really complete when the genetic composition of the two groups becomes so different that the chromosomes supplied by the male cannot match those supplied by the female and therefore an embryo, a new individual, cannot form. The change need not be large, and the genetic difference between one species and another may be subtle. All individuals are genetically unique, after all, and so the genetic composition of a species is at best a kind of average. The difference between one human and another may be greater than the difference between an 'average' human and an 'average' chimpanzee. Humans and chimpanzees are very similar genetically. Indeed, they are related to one another so closely that some scientists have proposed they be placed in the same genus or even that they be ranked as sub-species of the same species. Yet, slight though they may be, the genetic differences produce obvious and profound physical differences.

The genetic composition of a population changes over the generations. Chance mutations occur, altering bits of the genetic code. Such mutations are very common. They occur every day to most of us. In most cases they produce no outward effect. In its new, altered form the genetic material is useless. It is abandoned by the cell that contains it, or the cell itself dies. If it should be a sperm or egg cell that is affected, then that cell becomes incompatible, and cannot unite with its opposite number. Mating may occur, but conception does not. Where the change is such that conception is possible, in the great majority of cases the genetic change is lethal for the individual possessing it. It makes it impossible for the new embryo to survive and so it is lost. If the new individual is able to live, often it is only for a short time, so that death occurs before sexual maturity is reached. For whatever reason, the gene mutations that occur daily almost invariably fail to contribute anything useful, and the affected genes are rejected before they can be transmitted to a second generation. This is only what we should expect, for species of living organisms thrive only if they are well adapted to the conditions under which they live, and after many generations of natural selection there is likely to be little room for improvement.

Now and then, though, a change may occur that confers some benefit on the individual possessing it. Perhaps the environment has changed, so altering the 'rules of the game', and imposing a new kind of selection pressure. That individual becomes better able to live in its environment and because it is to this degree better equipped, it is likely to produce more offspring, all of whom inherit the desirable trait, although only some of them will express it. After a number of generations, the 'new gene' becomes established in the population at large, and then we may say that the genetic composition of the population as a whole has changed. Obviously, if the population moves, and so finds itself living under conditions different from those under which its ancestors lived, the more successful 'new genes' will be those that improve the adaptation of individuals to the new environment.

Genes can spread among a population only by breeding within the population. There is no way they can be acquired by individuals who are not descended from individuals possessing them. When a single population becomes divided reproductively, therefore, 'new genes' that appear in one of the groups cannot occur in the other, and little by little the two populations will drift apart from one another, becoming more and more different until the point is reached at which reproduction between them becomes genetically impossible. Since gene mutations occur frequently, it is only a matter of time before reproductive isolation, if it is total, and if there is pressure to select individuals who possess new qualities, will lead to the evolution of new species.

That, then, was the situation in which my martian ancestors found themselves, but in an extreme form. Among them the rate of mutation increased substantially. Ionizing radiation is one cause of gene mutations. During their journey to Mars they were exposed to higher levels of radiation than those they had experienced on Earth, and that increased the rate of mutation. On Mars itself the background radiation level is somewhat higher than the level on Earth, and that, too, increased the mutation rate. As I said, most mutations are lethal to themselves or to the organisms carrying them, but non-lethal mutations occur very occasionally, so that if the mutation rate increases, so does the rate at which non-lethal mutations occur and become fixed in the population.

Then again, the new martian environment was radically different from the old terran environment—from any terran environment. Mutations that were non-lethal and that actually helped humans to adapt to martian conditions became fixed in the population more rapidly than they would have done otherwise. That, too, accelerated the rate of change. I mentioned the initiation test that young men were required to pass, and the effect of our two epidemics of pneumonia.

We do not look much different from Terrans, but you can tell us apart. We move differently, as you would expect of people used to a different gravitational field, and our faces are a little different, too. Because we breathe a carbon dioxide-oxygen atmosphere, and have deep voices as a result, we use our speech muscles a little differently from Terrans, and although the characteristic is not inherited, after living for some years on Mars small changes in the development of facial and throat muscles alter slightly the shape of our faces. As we adjust to breathing martian air, we learn to breathe more efficiently, which gives us a better posture than many Earth-bound Terrans and encourages us to make use of our rib cages, as singers and actors have always been taught to do.

The important differences are not visible. We are rather more tolerant of radiation than are Terrans, both ionizing radiation and ultraviolet. We tan easily and are basically rather brown-skinned. We sweat less than Terrans do, and we pee less, but our urine contains our bodily wastes in a more concentrated form. This is a minor adaptation to rather dry conditions. We conserve water better, just as on

Earth elands in Africa use water more efficiently than the cattle to which they are related. We have not adapted in any special way to the martian atmosphere, because no such adaptation was possible or desirable. All humans adapt very quickly to reduced amounts of oxygen, but since we can have as much oxygen as we like indoors, and since we use breathing apparatus outdoors, no gain is to be made there. No gain, that is, except one. A Martian may live for just a little longer on the oxygen liberated from the soil should his or her breathing apparatus fail. That may be the result of genetic change, but it may also be a matter of training and familiarity. If your family has lived here for generations you know what to do in an emergency without pausing to think about it, and without panicking, which wastes oxygen. It may well be a cultural adaptation, like that of the terran Australian aboriginal peoples to life in their desert.

It is all hindsight, of course. Suspicions were aroused slowly. Martian-Terran marriages produced few children, and no grandchildren. It was a long time before anyone noticed this because, as I have said, such marriages were rare. Then, about fifteen (martian) years ago, the youngsters from one of the oldest settlements left home, moved right out to the periphery, and built themselves a new settlement in the far north, in Phlegra. They were radicals, naturally, and their radicalism insisted on the breaking down of social barriers. They felt it was wrong, indeed socially dangerous, that some of us should consider ourselves 'senior' to others. I remember the episode well, because I was almost as old as they were and was impressed by what I saw as their idealism. Anyway, they began to visit more recent settlements in Hecates Lacus.

I should explain, perhaps, that these regions are more or less due north of my own region, because as settlements

spread they hugged the equator, so that the periphery soon
came to lie along the northern and southern borders of the
inhabited regions. The Terrans welcomed them, and marriages followed, then children, but not many children—
which was remarked upon but was put down to the new
forms of behaviour the young people were trying to establish. It was not until the birth rate fell even further among
the second generation, so that the whole group were bound
to die out, that serious questions were asked. This could
not be explained as aberrant behaviour.

By this time more years had passed, and rather less than
two years ago all the people involved were examined and
their gene patterns compared. This led to a wider survey of
martian and terran groups until eventually the entire population of the planet had been surveyed. I do not mean that
every individual on the planet was examined, of course. It
was not necessary to go so far as that to produce what
martian scientists regarded as clear evidence for speciation,
based on a sex-linked genetic change. Martian women
cannot produce healthy, fertile offspring if the father is
terran, but terran women can bear the children of martian
fathers.

The scientists wrote their report, describing the history
of their investigations, their laboratory methods, the statis-
tical techniques they had used, and their conclusions. That
was the report I carried to Earth, where terran scientists
examined it, checked it, and confirmed it. It was, is, true.

I travelled to Earth suspecting, and returned knowing
that a gulf now separates me from my terran friends. We
are all human, but we are no longer the same kind of
human.

It is not a matter for regret in itself. As we explore and
colonize the universe it is bound to happen. Humans
inevitably must evolve into more and more divergent

forms. That is biology, nothing more. There is no point in regretting it, and quite a lot of point in welcoming it. There may be real improvements!

Yet it is disturbing, for on Mars both kinds of human must live together. Our society has developed its own ways of dealing with conflict, but I wonder whether it will remain so tolerant of a difference that is so profound? Will we 'discover' racialism all over again? Perhaps we will not. Perhaps we may avoid that trap, for although two races have emerged, they are free from the single most disruptive element that disfigured relations among the old, imaginary human 'races' back on Earth. There can be no interbreeding, although there can be infertile intermarriage.

Nor does it end here. With two races of humans in the solar system how certain may we be that Mars will not spawn more? Our self-sufficient settlements and accelerated mutation rate continue to provide ideal conditions for speciation. Will the day dawn when Mars is populated by dozens, scores, of human variants? Will it matter?

I speculate about matters that are beyond me. I ask questions to which there are no answers. It is one thing to discuss our evolution as a process over which we have little or no control, but quite another to try to imagine how our societies may adjust to the changes time forces upon them. We could take our evolution into our own hands, I dare say. We do this to some extent already, and have done for many, many years. We manipulate genes to eliminate those that are harmful to us, so that deformities can no longer be inherited. Could we cobble the two species back together again? Who can tell? Will we choose even to try?

If humans are to diverge further, perhaps they will do so elsewhere than on Mars or Earth. We have come this far, have settled a new world and made it our own. It is impossible for us not to gaze sometimes at the night sky and to wonder. Where will we go next?

There are no planets suitable for us. We could establish settlements of a kind on some solar system planets, as colonies have been established on the Moon, but there is no other planet we could transform as life transformed Earth and as, with the help of some chemicals, algae, bacteria, and then plants, we have transformed Mars. Venus is the candidate human legend prefers, but the transformation of Venus would be a very different proposition from the transformation of Mars. We have considered Venus, and rejected it. Nothing had changed that would make us revise our view.

Inward from Venus there lies Mercury, but is has no atmosphere at all and is a most inhospitable place. Outward from Mars lies the asteroid belt, containing no body large enough to be colonized in the conventional sense.

Beyond the asteroid belt, and a long way beyond them, lie the giant planets and their satellites. They, too, are in chemical and thermal equilibrium and, to complicate matters, they are so far from the Sun that there is no outside energy source living things might exploit. Photosynthesis, for example, would be impossible in the remoter reaches of the solar system.

Now and then someone suggests that we go to live on Titan, a satellite of Saturn rather less than half the size of Earth, with an atmosphere, weather, and oceans. It is not very promising. The oceans apparently consist of liquid methane. No doubt the night sky on Titan presents a splendid view of the giant Saturn, but life at the surface would be far from comfortable and there is no way I can imagine of establishing living organisms there.

Should we conclude, then, that our search for new

places to live must now take us away from the solar system altogether? The answer to that must be yes, and no. It is true that the system offers no new planet for us to colonize, so if it is another planet we seek we must find another star. It is not true, not quite, that henceforth humans must confine themselves to Earth, Mars, and the Moon. Mars, too, has satellites, and in recent years there have been Martians who have eyed them with interest.

Perhaps we could make use of Phobos and Deimos. They are small, but if they were hollowed out, might they not be converted into the most magnificently large space-ships? Attach some means of propulsion to them and they could sail the galaxy for as many human generations as it might take in their search for new worlds, and during their voyage they would be small, self-contained worlds in themselves. Mars would not miss them. They are too small to provide us with light at night, and much too small to exert any kind of tidal effect on our world—even if we had enough surface water to be affected by tides.

When you remember that it costs much less energy to launch a payload into orbit from Mars than it does from Earth, the conversion of existing satellites to new uses begins to look attractive. The basic structure exists already, so that once it had been hollowed we would need only to furnish and equip it.

We have come so far. Humans have left Earth. They have made the really 'big jump' by leaving their home planet and learning to live on another. It is impossible that they should not think of going further, of continuing the adventure, just as on Earth their discovery of new lands never satisfied them, but served only to stimulate the search for more new lands beyond the horizon.

Is it possible? Could humans really travel to other stars? Could they traverse the galaxy? They could, provided they

abandon three assumptions. They must give up any hope
of travelling to another star and then returning to the world
from which they started. That would be quite impossible
because of the great distances involved. Having accepted
that limitation, they must also give up hope even of
maintaining communication with their home planet for
very long. They might be able to send messages, or even
to receive them, but once they had covered a distance of a
few light-weeks, conversations would become impossibly
slow, because radio signals travel at the speed of light.
People would soon give up if, at a distance of even two
light-weeks, they had to wait two weeks for their question
to reach its destination, and two more weeks for a reply.
Finally, having abandoned hope of returning home, and of
maintaining close contact with home, they must give up
hope of ever arriving. They must cover distances that are
truly vast, and the human lifespan is short. A community
would depart, but if ever it arrived anywhere the people
who arrived would be related as closely to the original
pioneers as I am, say, to the people of the European
Middle Ages. I am descended from them, because I must
be, but I know of them only what I read in books.

Accept the limitations, and perhaps star travel is possible. Build a ship, devise it in such a way that it will continue
to provide for its crew for centuries and perhaps for
millenia, provide it with a means of propulsion, and
accelerate it. Continue to accelerate it at a constant rate
and in space it will not be long before it is travelling at
half the speed of light, or even faster. Then shut down the
engines and cruise, and you will be able to cruise for ever
more.

We think, and we dream. We think of our past, the past we
shared with Terrans, and of the recent past, that is truly

our own. We try to comprehend the change that has taken place in us, to know in our hearts what it will mean. Soon my report must be published, must become the property of all, and everyone will share the sudden loneliness I feel. An era in human history has ended, and an era has begun, and the split human race henceforth must pursue more or less parallel paths. More? Or less?

We cannot know, therefore we dream. We dream of the future, of our future, and having looked behind us to another planet, in our dreams we cannot help looking ahead to more explorations. One day, perhaps, some remote descendant of mine will walk by the light of another sun on a world made by people with the help of chloro-fluoro-carbon compounds, algae, bacteria, lichens, ferns, mosses. Dare I think that person as human, or even hominid?

Dreams are no more than dreams. Can they be translated into deeds? Are our ideas about the conversion of Phobos and Deimos, of voyages lasting countless generations, mere fantasy? Perhaps they are. Perhaps we are bound to our two planets. Yet it is not so long since most sensible poeple would have dismissed out of hand any suggestion not only that they leave Earth and make homes for themselves on Mars, but that they do so in the very near future. Was that pure fantasy? They were not alone. The list is long of those who have believed the future to be impossible.

Our dreams are no more than dreams, but dreams can come true.

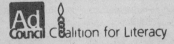